Fundamentals of Quantum Mechanics

Quantum mechanics has evolved from a subject of study in pure physics to one with a wide range of applications in many diverse fields. The basic concepts of quantum mechanics are explained in this book in a concise and easy-to-read manner, leading toward applications in solid state electronics and modern optics. Following a logical sequence, the book is focused on the key ideas and is conceptually and mathematically self-contained. The fundamental principles of quantum mechanics are illustrated by showing their application to systems such as the hydrogen atom, multi-electron ions and atoms, the formation of simple organic molecules and crystalline solids of practical importance. It leads on from these basic concepts to discuss some of the most important applications in modern semiconductor electronics and optics.

Containing many homework problems, the book is suitable for senior-level undergraduate and graduate level students in electrical engineering, materials science, and applied physics and chemistry.

C. L. Tang is the Spencer T. Olin Professor of Engineering at Cornell University, Ithaca, NY. His research interest has been in quantum electronics, nonlinear optics, femtosecond optics and ultrafast process in molecules and semiconductors, and he has published extensively in these fields. He is a Fellow of the IEEE, the Optical Society of America, and the Americal Physical Society, and is a member of the US National Academy of Engineering. He was the winner of the Charles H. Townes Award of the Optical Society of America in 1996.

Fundamentals of Quantum Mechanics

For Solid State Electronics and Optics

C. L. TANG
Cornell University, Ithaca, NY

CAMBRIDGE
UNIVERSITY PRESS

CAMBRIDGE UNIVERSITY PRESS

Cambridge, New York, Melbourne, Madrid, Cape Town, Singapore, São Paulo

CAMBRIDGE UNIVERSITY PRESS

The Edinburgh Building, Cambridge CB2 2RU, UK

Published in the United States of America by Cambridge University Press, New York

www.cambridge.org
Information on this title: www.cambridge.org/9780521829526

First published 2005

Printed in the United Kingdom at the University Press, Cambridge

A catalog record for this book is available from the British Library

ISBN-13 978-0-521-82952-6 hardback
ISBN-10 0-521-82952-6 hardback

To
Louise

Contents

Preface

Quantum mechanics has evolved from a subject of study in pure physics to one with a vast range of applications in many diverse fields. Some of its most important applications are in modern solid state electronics and optics. As such, it is now a part of the required undergraduate curriculum of more and more electrical engineering, materials science, and applied physics schools. This book is based on the lecture notes that I have developed over the years teaching introductory quantum mechanics to students at the senior/first year graduate school level whose interest is primarily in applications in solid state electronics and modern optics.

There are many excellent introductory text books on quantum mechanics for students majoring in physics or chemistry that emphasize atomic and nuclear physics for the former and molecular and chemical physics for the latter. Often, the approach is to begin from a historic perspective, recounting some of the experimental observations that could not be explained on the basis of the principles of classical mechanics and electrodynamics, followed by descriptions of various early attempts at developing a set of new principles that could explain these 'anomalies.' It is a good way to show the students the historical thinking that led to the discovery and formulation of the basic principles of quantum mechanics. This might have been a reasonable approach in the first half of the twentieth century when it was an interesting story to be told and people still needed to be convinced of its validity and utility. Most students today know that quantum theory is now well established and important. What they want to know is not how to reinvent quantum mechanics, but what the basic principles are concisely and how they are used in applications in atomic, molecular, and solid state physics. For electronics, materials science, and applied physics students in particular, they need to see, above all, how quantum mechanics forms the foundations of modern semiconductor electronics and optics. To meet this need is then the primary goal of this introductory text/reference book, for such students and for those who did not have any quantum mechanics in their earlier days as an undergraduate student but wish now to learn the subject on their own.

This book is not encyclopedic in nature but is focused on the key concepts and results. Hopefully it makes sense pedagogically. As a textbook, it is conceptually and mathematically self-contained in the sense that all the results are derived, or derivable, from first principles, based on the material presented in the book in a logical order without excessive reliance on reference sources. The emphasis is on concise physical explanations, complemented by rigorous mathematical demonstrations, of how things work and why they work the way they do.

A brief introduction is given in **Chapter 1** on how one goes about formulating and solving problems on the atomic and subatomic scale. This is followed in **Chapter 2** by a concise description of the basic postulates of quantum mechanics and the terminology and mathematical tools that one will need for the rest of the book. This part of the book by necessity tends to be on the abstract side and might appear to be a little formal to some of the beginning students. *It is not necessary to master all the mathematical details and complications at this stage.* For organizational reasons, I feel that it is better to collect all this information at one place at the beginning so that the flow of thoughts and the discussions of the main subject matter will not be repeatedly interrupted later on by the need to introduce the language and tools needed.

The basic principles of quantum mechanics are then applied to a number of simple prototype problems in **Chapters 3–5** that help to clarify the basic concepts and as a preparation for discussing the more realistic physical problems of interest in later chapters. **Section 5.4** on photons is a discussion of the application of the basic theory of harmonic oscillators to radiation oscillators. It gives the basic rules of quantization of electromagnetic fields and discusses the historically important problem of black-body radiation and the more recently developed quantum theory of coherent optical states. For an introductory course on quantum mechanics, this material can perhaps be skipped.

Chapters 6 and **7** deal with the hydrogenic and multi-electron atoms and ions. Since the emphasis of this book is not on atomic spectroscopy, some of the mathematical details that can be found in many of the excellent books on atomic physics are not repeated in this book, except for the key concepts and results. These chapters form the foundations of the subsequent discussions in **Chapter 8** on the important topics of time-dependent perturbation theory and the interaction of radiation with matter. It naturally leads to Einstein's theory of resonant absorption and emission of radiation by atoms. One of its most important progeny is the ubiquitous optical marvel known as the LASER (Light Amplification by Stimulated Emission of Radiation).

From the hydrogenic and multi-electron atoms, we move on to the increasingly more complicated world of molecules and solids in **Chapter 9**. The increased complexity of the physical systems requires more sophisticated approximation procedures to deal with the related mathematical problems. The basic concept and methodology of time-independent perturbation theory is introduced and applied to covalent-bonded diatomic and simple organic molecules. Crystalline solids are in some sense giant molecules with periodic lattice structures. Of particular interest are the sp^3-bonded elemental and compound semiconductors of diamond and zincblende structures.

Some of the most important applications of quantum mechanics are in semiconductor physics and technology based on the properties of charge-carriers in periodic lattices of ions. Basic concepts and results on the electronic properties of semiconductors are discussed in **Chapter 10**. The molecular-orbital picture and the nearly-free-electron model of the origin of the conduction and valence bands in semiconductors based on the powerful Bloch theorem are developed. From these follow the commonly used concepts and parameters to describe the dynamics of charge-carriers in semiconductors, culminating finally in one of the most important building blocks of modern electronic and optical devices: the p–n junction.

For applications involving macroscopic samples of many particles, the basic quantum theory for single-particle systems must be generalized to allow for the situation where the quantum states of the particles in the sample are not all known precisely. For a uniform sample of the same kind of particles in a statistical distribution over all possible states, the simplest approach is to use the density-matrix formalism. The basic concept and properties of the density operator or the density matrix and their equations of motion are introduced in **Chapter 11**. This chapter, and the book, conclude with some examples of applications of this basic approach to a number of linear and nonlinear, static and dynamic, optical problems. For an introductory course on quantum mechanics, this chapter could perhaps be omitted also.

While there might have been, and may still be in the minds of some, doubts about the basis of quantum mechanics on philosophical grounds, there is no ambiguity and no doubt on the applications level. The rules are clear, precise, and all-encompassing, and the predictions and quantitative results are always correct and accurate without exception. It is true, however, that at times it is difficult to penetrate through the mathematical underpinnings of quantum mechanics to the physical reality of the subject. I hope that the material presented and the insights offered in this book will help pave the way to overcoming the inherent difficulties of the subject for some. It is hoped, above all, that the students will find quantum mechanics a fascinating subject to study, not a subject to be avoided.

I am grateful for the opportunities that I have had to work with the students and many of my colleagues in the research community over the years to advance my own understanding of the subject. I would like to thank, in particular, Joe Ballantyne, Chris Flytzanis, Clif Pollck, Peter Powers, Hermann Statz, Frank Wise, and Boris Zeldovich for their insightful comments and suggestions on improving the presentation of the material and precision of the wording. Finally, without the numerous questions and puzzling stares from the generations of students who have passed through my classes and research laboratory, I would have been at a loss to know what to write about.

A note on the unit system: to facilitate comparison with classic physics literature on quantum mechanics, the unrationalized cgs Gaussian unit system is used in this book unless otherwise stated explicitly.

1 Classical mechanics vs. quantum mechanics

What is quantum mechanics and what does it do?

In very general terms, the basic problem that both classical Newtonian mechanics and quantum mechanics seek to address can be stated very simply: *if the state of a dynamic system is known initially and something is done to it, how will the state of the system change with time in response?*

In this chapter, we will give a brief overview of, first, how Newtonian mechanics goes about solving the problem for systems in the macroscopic world and, then, how quantum mechanics does it for systems on the atomic and subatomic scale. We will see qualitatively what the differences and similarities of the two schemes are and what the domain of applicability of each is.

1.1 Brief overview of classical mechanics

To answer the question posed above systematically, we must first give a more rigorous formulation of the problem and introduce the special language and terminology (in double quotation marks) that will be used in subsequent discussions. For the macroscopic world, common sense tells us that, to begin with, we should identify the "system" that we are dealing with in terms of a set of "static properties" that do not change with time in the context of the problem. For example, the mass of an object might be a static property. The change in the "state" of the system is characterized by a set of "dynamic variables." Knowing the initial state of the system means that we can specify the "initial conditions of these dynamic variables." What is done to the system is represented by the "actions" on the system. How the state of the system changes under the prescribed actions is then described by how the dynamic variables change with time. This means that there must be an "equation of motion" that governs the time-dependence of the state of the system. The mathematical solution of the equation of motion for the dynamic variables of the system will then tell us precisely the state of the system at a later time $t > 0$; that is to say, everything about what happens to the system after something is done to it.

For definiteness, let us start with the simplest possible "system": a single particle, or a point system, that is characterized by a single static property, its mass m. We assume that its motion is limited to a one-dimensional linear space (1-D, coordinate axis x, for example). According to Newtonian mechanics, the state of the particle at any time t is completely specified in terms of the numerical values of its position $x(t)$ and velocity $v_x(t)$, which is the rate of change of its position with respect to time, or $v_x(t) = \mathrm{d}x(t)/\mathrm{d}t$. All the other dynamic properties, such as linear momentum $p_x(t) = mv_x$, kinetic energy $T = (mv_x^2)/2$, potential energy $V(x)$, total energy $E = (T + V)$, etc. of this system depend only on x and v_x. "The state of the system is known initially" means that the numerical values of $x(0)$ and $v_x(0)$ are given. The key concept of Newtonian mechanics is that the action on the particle can be specified in terms of a "force", F_x, acting on the particle, and this force is proportional to the acceleration, $a_x = \mathrm{d}^2x/\mathrm{d}t^2$, where the proportionality constant is the mass, m, of the particle, or

$$F_x = ma_x = m\frac{\mathrm{d}^2x}{\mathrm{d}t^2}. \tag{1.1}$$

This means that once the force acting on a particle of known mass is specified, the second derivative of its position with respect to time, or the acceleration, is known from (1.1). With the acceleration known, one will know the numerical value of $v_x(t)$ at all times by simple integration. By further integrating $v_x(t)$, one will then also know the numerical value of $x(t)$, and hence what happens to the particle for all times. Thus, if the initial conditions on x and v_x are given and the action, or the force, on the particle is specified, one can always predict the state of the particle for all times, and the initially posed problem is solved.

The crucial point is that, because the state of the particle is specified by x and its first time-derivative v_x to begin with, in order to know how x and v_x change with time, one only has to know the second derivative of x with respect to time, or specify the force. This is a basic concept in calculus which was, in fact, invented by Newton to deal with the problems in mechanics.

A more complicated dynamic system is composed of many constituent parts, and its motion is not necessarily limited to any one-dimensional space. Nevertheless, no matter how complicated the system and the actions on the system are, the dynamics of the system can, in principle, be understood or predicted on the basis of these same principles. In the macroscopic world, the validity of these principles can be tested experimentally by direct measurements. Indeed, they have been verified in countless cases. The principles of Newtonian mechanics, therefore, describe the "*laws of Nature*" in the macroscopic world.

1.2   Overview of quantum mechanics

What about the world on the atomic and subatomic scale? A number of fundamental difficulties, both experimental and logical, immediately arise when trying to extend the principles of Newtonian mechanics to the atomic and subatomic scale. For example, measurements on atomic or subatomic particles carried out in the macroscopic world in general give results that are statistical averages over an ensemble of a large number of similarly prepared particles, not precise results on any particular particle. Also, the resolution needed to quantify or specify the properties of individual systems on the atomic and subatomic scale is generally many orders of magnitude finer than the scales and accuracy of any measurement process in the macroscopic world. This makes it difficult to compare the predictions of theory with direct measurements for specific atomic or subatomic systems. Without clear direct experimental evidence, there is no a priori reason to expect that it is *always* possible to specify the state of an atomic or subatomic particle at any particular time in terms of a set of simultaneously precisely measurable parameters, such as the position and velocity of the particle, as in the macroscopic world. The whole formulation based on the deterministic principles of Newtonian mechanics of the basic problem posed at the beginning of this discussion based on simultaneous precisely measurable position and velocity of a particular particle is, therefore, questionable. Indeed, while Newtonian mechanics had been firmly established as a valid theory for explaining the behaviors of all kinds of dynamic systems in the macroscopic world, experimental anomalies that could not be explained by such a theory were also found in the early part of the twentieth century. Attempts to explain these anomalies led to the development of quantum theory, which is a totally new way of dealing with the problems of mechanics and electrodynamics in the atomic and subatomic world.

A brief overview of the general approach of the theory in contrast to classical Newtonian mechanics is given here. All the assertions made in this brief overview will be explained and justified in detail in the following chapters. The purpose of the *qualitative* discussion in this chapter is simply to give an indication of the things to come, not a complete picture. A more formal description of the basic postulates and methodology of quantum mechanics will be given in the following chapter.

To begin with, according to quantum mechanics, the "state" of a system on the atomic and subatomic scale is not characterized by a set of dynamic variables each with a specific numerical value. Instead, it is completely specified by a "state function." The dynamics of the system is described by the time dependence of this state function. The relationship between this state function and various physical properties of the dynamic system that can be measured in the macroscopic world is also not as direct as in Newtonian mechanics, as will be clarified later.

The state function is a function of a set of chosen variables, called "canonic variables," of the system under study. For definiteness, let us consider again, for example, the case of a particle of mass m constrained to move in a linear space along the x axis. The state function, which is usually designated by the arbitrarily chosen symbol Ψ, is a function of x. That is, the state of the particle is specified by the functional dependence of the state function $\Psi(x)$ on the canonic variable x, which is the "possible position" of the particle. It is not specified by any particular values of x and v_x as in Newtonian mechanics. How the state of the particle changes with time is specified by $\Psi(x, t)$, or how $\Psi(x)$ changes explicitly with time, t. $\Psi(x, t)$ is often also referred to as the "wave function" of the particle, because it often has properties similar to those of a wave, even though it is supposed to describe the state of a "particle," as will be shown later.

The state function can also be expressed alternatively as a function of another canonic variable "conjugate" to the position coordinate of the system, the linear momentum of the particle p_x, or $\Psi(p_x, t)$. The basic problem of the dynamics of the particle can be formulated in either equivalent form, or in either "representation." If the form $\Psi(x, t)$ is used, it is said to be in the "Schrödinger representation," in honor of one of the founders of quantum mechanics. If the form $\Psi(p_x, t)$ is used, it is in the "momentum representation." That the same state function can be expressed as a function of different variables corresponding to different representations is analogous to the situation in classical electromagnetic theory where a time-dependent electrical signal can be expressed either as a function of time, $\varepsilon(t)$, or in terms of its angular-frequency spectrum, $\varepsilon(\omega)$, in the Fourier-transform representation. There is a unique relationship between $\Psi(x, t)$ and $\Psi(p_x, t)$, much as that between $\varepsilon(t)$ and $\varepsilon(\omega)$. Either representation will eventually lead to the same results for experimentally measurable properties, or the "observables," of the system. Thus, as far as interpreting experimental results goes, it makes no difference which representation is used. The choice is generally dictated by the context of the problem or mathematical expediency. Most of the introductory literature on the quantum theory of electronic and optical devices tends to be based on the Schrödinger representation. That is what will be mostly used in this book also.

The "statistical," or probabilistic, nature of the measurement process on the atomic and subatomic scale is imbedded in the physical interpretation of the state function. For example, the wave function $\Psi(x, t)$ is in general a complex function of x and t, meaning it is a phasor of the form $\Psi = |\Psi|\, e^{i\phi}$ with an amplitude $|\Psi|$ and a phase ϕ. The magnitude of the wave function, $|\Psi(x, t)|$, gives statistical information on the results of measurement of the position of the particle. More specifically, "the particle" in quantum mechanics actually means a statistical "ensemble," or collection, of particles all in the same state, Ψ, for example. $|\Psi(x, t)|^2 \mathrm{d}x$ is then interpreted as the probability of finding a particle in the ensemble in the spatial range from x to $x + \mathrm{d}x$ at the time t. Unlike in Newtonian mechanics, we cannot speak of the precise position of a specific atomic or subatomic particle in a statistical ensemble of particles. The experimentally measured position must be viewed as an "expectation value," or the average value, of the probable position of the particle. An explanation of the precise meanings of these statements will be given in the following chapters.

The physical interpretation of the phase of the wave function is more subtle. It endows the particle with the "duality" of wave properties, as will be discussed later. The statistical interpretation of the measurement process and the wave–particle duality of the dynamic system represent fundamental philosophical differences between the quantum mechanical and Newtonian descriptions of "dynamic systems."

For the equation of motion in quantum mechanics, we need to specify the "action" on the system. In Newtonian mechanics, the action is specified in terms of the force acting on the system. Since the force is equal to the rate of decrease of the potential energy with the position of the system, or $\vec{F} = -\nabla V(\vec{r})$, the action on the system can be specified either in terms of the force acting on the system or the potential energy of the particle as a function of position $V(\vec{r})$. In quantum mechanics, the action on the dynamic system is generally specified by a physically "observable" property corresponding to the "potential energy operator," say $\hat{V}(\vec{r})$, as a function of the position of the system. For example, in the one-dimensional single-particle problem, $\hat{V}$ in the Schrödinger representation is a function of the variable x, or $\hat{V}(x)$. Since the position of a particle in general does not have a unique value in quantum mechanics, the important point is that $\hat{V}(x)$ gives the functional relationship between $\hat{V}$ and the position variable x. The force acting on the system is simply the negative of the gradient of the potential with respect to x; therefore, the two represent the same physical action on the system. Physically, $\hat{V}(x)$ gives, for example, the direction in which the particle position must change in order to lower its potential energy; it is, therefore, a perfectly reasonable way to specify the action on the particle.

In general, all dynamic properties are represented by "operators" that are functions of x and $\hat{p}_x$. As a matter of notation, a 'hat $\wedge$' over a symbol in the language of quantum theory indicates that the symbol is mathematically an "operator," which in the Schrödinger representation can be a function of x and/or a differential operator involving x. For example, the operator representing the linear momentum, $\hat{p}_x$, in the Schrödinger representation is represented by an operator that is proportional to the first derivative with respect to x:

$$\hat{p}_x = -i\,\hbar \frac{\partial}{\partial x}, \tag{1.2}$$

where $\hbar$ is the Planck's constant h divided by 2π. h is one of the fundamental constants in quantum mechanics and has the numerical value $h = 6.626 \times 10^{-27}$ erg-s. The reason for this peculiar equation, (1.2), is not obvious at this point. It is related to one of the basic "postulates" of quantum mechanics and one of its implications is the all-important "Heisenberg's uncertainty principle," as will be discussed in detail in later chapters.

The total energy of the system is generally referred to as the "Hamiltonian," and usually represented by the symbol $\hat{H}$, of the system. It is the sum of the kinetic energy and the potential energy of the system as in Newtonian mechanics:

$$\hat{H} = \frac{\hat{p}_x^2}{2m} + \hat{V}(x) = -\frac{\hbar^2}{2m}\frac{\partial^2}{\partial x^2} + \hat{V}(x), \tag{1.3}$$

with the help of Eq. (1.2). The action on the system is, therefore, contained in the Hamiltonian through its dependence on $\hat{V}$.

The total energy, or the Hamiltonian, plays an essential role in the equation of motion dealing with the dynamics of quantum systems. *Because the state of the dynamic system in quantum mechanics is completely specified by the state function, it is only necessary to know its first time-derivative, $\frac{\partial \Psi}{\partial t}$, in order to predict how Ψ will vary with time, starting with the initial condition on Ψ.* The key equation of motion as postulated by Schrödinger is that the time-rate of change of the state function is proportional to the Hamiltonian "operating" on the state function:

$$i\hbar\frac{\partial \Psi}{\partial t} = \hat{H}\Psi. \tag{1.4}$$

In the Schrödinger representation for the one-dimensional single particle system, for example, it is a partial differential equation:

$$i\hbar\frac{\partial \Psi}{\partial t} = \left[-\frac{\hbar^2}{2m}\frac{\partial^2}{\partial x^2} + \hat{V}(x)\right]\Psi, \tag{1.5}$$

by substituting Eq. (1.3) into Eq. (1.4). The time-dependent Schrödinger's equation, Eq. (1.4), or more often its explicit form Eq. (1.5), is the basic equation of motion in quantum mechanics that we will see again and again later in applications. Solution of Schrödinger's equation will then describe completely the dynamics of the system.

The fact that the basic equation of motion in quantum mechanics involves only the *first* time-derivative of something while the corresponding equation in Newtonian mechanics involves the *second* time-derivative of some key variable is a very interesting and significant difference. It is a necessary consequence of the fundamental difference in how the "state of a dynamic system" is specified in the two approaches to begin with. It also leads to the crucial difference in how the action on the system comes into play in the equations of motion: the total energy, $\hat{H}$, in the former case, in contrast to the force, $\vec{F}$, in the latter case.

Schrödinger's equation, (1.4), in quantum mechanics is analogous to Newton's equation of motion, Eq. (1.1), in classical mechanics. It is one of the key postulates that unlocks the wonders of the atomic and subatomic world in quantum mechanics. It has been verified with great precision in numerous experiments without exception. It can, therefore, be viewed as a *law of Nature* just as Newton's equation – 'F equals m a' – for the macroscopic world.

The problem is now reduced to a purely mathematical one. Once the initial condition $\Psi(x, t = 0)$ and the action on the system are given, the solution of the Schrödinger equation gives the state of the system at any time t. Knowing $\Psi(x, t)$ at any time t also means that we can find the expectation values of all the operators corresponding to the dynamic properties of the system. Exactly how that is done mathematically will be described in detail in the following chapters. Since the state of the system is completely specified by the state function, the time dependent state function $\Psi(\vec{r}, t)$ contains all the information on the dynamics of the system that can be obtained by experimental observations. This is how the problem is formulated and solved according to the principles of quantum mechanics.

Further reading

For further studies at a more advanced level of the topics discussed in this and the following chapters of this book, we recommend the following.

On fundamentals of quantum mechanics

Bethe and Jackiw (1986); Bohm (1951); Cohen-Tannoudji, Diu and Laloë (1977); Dirac (1947).

On quantum theory of radiation

Glauber (1963); Heitler (1954).

On generalized angular momentum

Edmonds (1957); Rose (1956).

On atomic spectra and atomic structure

Condon and Shortley (1963); Herzberg (1944).

On molecules and molecular-orbital theory

Ballhausen and Gray (1964); Coulson (1961); Gray (1973); Pauling (1967).

On lasers and photonics

Siegman (1986); Shen (1984); Yariv (1989).

On solid state physics and semiconductor electronics

Kittel (1996); Smith (1964); Streetman (1995).

2 Basic postulates and mathematical tools

Basic scientific theories usually start with a set of hypotheses or "postulates." There is generally no logical reason, apart from internal consistency, that can be given to justify such postulates absolutely. They come from 'revelations' in the minds of 'geniuses,' most likely with hints from Nature based on extensive careful observations. Their general validity can only be established through experimental verification. If numerous rigorously derived logical consequences of a very small set of postulates all agree with experimental observations without exception, one is inclined to accept these postulates as correct descriptions of the laws of Nature and use them confidently to explain and predict other natural phenomena. Quantum mechanics is no exception. It is based on a few postulates. For the purpose of the present discussion, we begin with three basic postulates involving: the "state functions," "operators," and "equations of motion."

In this chapter, this set of basic postulates and some of the corollaries and related definitions of terms are introduced and discussed. We will first simply state these postulates and introduce some of the related mathematical tools and concepts that are needed to arrive at their logical consequences later. To those who have not been exposed to the subject of quantum mechanics before, each of these postulates taken by itself may appear puzzling and meaningless at first. It should be borne in mind, however, that it is the collection of these postulates as a whole that forms the foundations of quantum mechanics. The full interpretation, and the power and glory, of these postulates will only be revealed gradually as they are successfully applied to more realistic and increasingly complicated physical problems in later chapters.

2.1 State functions (Postulate 1)

The first postulate states that *the state of a dynamic system is completely specified by a state function.*

Even without a clear definition of what a state function is, this simple postulate already makes a specific claim: there *exists* an abstract state function that contains *all* the information about the state of the dynamic system. For this statement to have meaning, we must obviously provide a clear physical interpretation of the state function, and specify its mathematical properties. We must also give a prescription of how quantitative information is to be extracted from the state function and compared with experimental results.

The state function, which is often designated by a symbol such as Ψ, is in general a complex function (meaning a phasor, $|\Psi|\,e^{i\phi}$, with an amplitude and a phase). In terms of the motion of a single particle in a linear space (coordinate x), for example, $|\Psi|$ and ϕ in the Schrödinger representation are functions of the canonical variable x.

A fundamental distinction between classical mechanics and quantum mechanics is that, in classical mechanics, the state of the dynamic system is completely specified by the position and velocity of each constituent part (or particle) of the system. This presumes that the position and velocity of a particle can, at least in principle, be measured and specified precisely at each instant of time. The position and velocity of the particle at one instant of time are completely determined by the position and velocity of the particle at a previous instant. It is deterministic. That one can specify the state of a particle in the macroscopic world in this way is intuitively obvious, because one can see and touch such a particle. It is intuitively obvious that it is possible to measure its position and velocity simultaneously. And, if two particles are not at the same place or not moving with the same velocity, they are obviously not in the same state.

What about in the world on the atomic and subatomic scale where we cannot see or touch any particle directly? There is no assurance that our intuition on how things work in our world can be extrapolated to a much smaller world in which we have no direct sensorial experience. Indeed, in quantum mechanics, no a priori assumption is made about the possibility of measuring or specifying precisely the position and the velocity of the particle at the same time. In fact, as will be discussed in more detail later, according to "Heisenberg's uncertainty principle," it is decidedly not possible to have complete simultaneous knowledge of the two; a complete formulation of this principle will be given in connection with Postulate 2 in Section 2.2 below. Furthermore, quantum mechanics does not presume that measurement of the position of a particle will necessarily yield a particular value of x predictably. Knowing the particle is in the state Ψ, the most specific information on the position of the particle that one can hope to get by any possible means of measurement is that the probability of getting the value x_1 relative to that of getting the value x_2 is $|\Psi(x_1)|^2$: $|\Psi(x_2)|^2$. In other words, the physical interpretation of the amplitude of the state function is that $|\Psi(x)|^2\mathrm{d}x$ is, in the language of probability theory, proportional to the probability of finding the particle in the range from x to $x + \mathrm{d}x$ in any measurement of the position of the particle. If it is known for certain that there is one particle in the spatial range from $x = 0$ to $x = L$, then the probability distribution function $|\Psi(x)|^2$ integrated over this range must be equal to 1 and the wave function is said to be "normalized":

$$1 = \int_0^L \Psi(x)^* \Psi(x)\mathrm{d}x = \int_0^L |\Psi(x)|^2 \mathrm{d}x. \tag{2.1}$$

If the wave function is normalized, the absolute value of the probability of finding the particle in the range from x to $x + \mathrm{d}x$ is $|\Psi(x)|^2\mathrm{d}x$. Accordingly, there is also an average value, $\langle x \rangle_\Psi$, of the position of the particle in the state Ψ, which is called the "expectation value" of the position of the particle. It is an ordinary number given by:

$$\langle x \rangle_\Psi = \int_0^L \Psi^*(x)\, x\, \Psi(x)\, \mathrm{d}x = \int_0^L x\, |\Psi(x)|^2 \mathrm{d}x. \tag{2.2}$$

A "mean square deviation," Δx^2, from the average of the probable position of the particle can also be defined:

$$\Delta x^2 = \int_0^L \Psi(x)^* (x - \langle x \rangle_\Psi)^2 \Psi(x) \mathrm{d}x = \int_0^L (x - \langle x \rangle_\Psi)^2 |\Psi(x)|^2 \mathrm{d}x, \tag{2.3}$$

which gives a measure of the spread of the probability distribution function, $|\Psi(x)|^2$, of the position around the average value. In the language of quantum mechanics, $\Delta x \equiv \sqrt{\Delta x^2}$ as defined in (2.3) is called the "uncertainty" in the position x of the particle when it is in the state $\Psi(x)$. The definitions of the "average value" and the "mean square deviation," or "uncertainty," can also be generalized to any function of x, such as *any operator* in the Schrödinger representation, as will be discussed in Section 2.3.

A more detailed explanation of the above probabilistic interpretation of the amplitude of the state function is in order at this point. "$|\Psi(x)|^2$ is the probability distribution function of the position of the particle" implies the following. If there are a large number of particles all in the same state Ψ in a statistical ensemble and similar measurement of the position of the particles is made on each of the particles in the ensemble, the result of the measurements is that the ratio of the number of times a particle is found in the range from x to $x + \mathrm{d}x$, N_x, to the total number of measurements, N, is equal to $|\Psi(x)|^2 \mathrm{d}x$. Stating it in another way, the number of times a particle is found in the differential range from x_1 to $x_1 + \mathrm{d}x$ to that in the range from x_2 to $x_2 + \mathrm{d}x$ is in the ratio of $N_{x_1}: N_{x_2} = |\Psi(x_1)|^2 : |\Psi(x_2)|^2$. The expectation value of the position of the particle, $\langle x \rangle_\Psi$, is the average of the measured positions of the particles:

$$\langle x \rangle_\Psi = x_1 \frac{N_{x_1}}{N} + x_2 \frac{N_{x_2}}{N} + x_3 \frac{N_{x_3}}{N} + \cdots = \int_0^L x\, |\Psi(x)|^2 \mathrm{d}x,$$

as given by Eq. (2.2). The uncertainty, Δx, is the spread of the measured positions around the average value:

$$\Delta x^2 = (x_1 - \langle x \rangle_\Psi)^2 \frac{N_{x_1}}{N} + (x_2 - \langle x \rangle_\Psi)^2 \frac{N_{x_2}}{N}$$
$$+ (x_3 - \langle x \rangle_\Psi)^2 \frac{N_{x_3}}{N} + \cdots$$
$$= \int_0^L (x - \langle x \rangle_\Psi)^2 |\Psi(x)|^2 \mathrm{d}x,$$

as given by Eq. (2.3).

The essence of the discussion so far is that the relationship between the physically measurable properties of a dynamic system and the state function of the system in quantum mechanics is probabilistic to begin with. The implication is that the prediction of the future course of the dynamics of the system in terms of physically measurable properties is, according to quantum mechanics, necessarily probabilistic, not deterministic, even though the time evolution of the state function itself is determined uniquely by its initial condition according to Schrödinger's equation, as we shall see.

It is also assumed as a part of Postulate 1 that the state function satisfies the "principle of superposition," meaning the linear combination of two state functions is also a possible state function:

$$\Psi = a_1 \Psi_1 + a_2 \Psi_2, \tag{2.4}$$

where a_1 and a_2 are, in general, complex numbers (with real and imaginary parts). This simple property has profound mathematical and physical implications, as will be seen later.

The physical significance of the phase, ϕ, of a state function, Ψ, is indirect and more subtle. In addition to its x-dependence, the phase factor also gives the explicit time-dependence of the wave function, as will be shown later in connection with the solution of Schrödinger's equation. It is, therefore, of fundamental importance to the understanding of the dynamics of atomic and subatomic particles.

The following example making use of the superposition principle may help to illustrate the physical significance of this phase factor. Suppose each particle in the state Ψ in the statistical ensemble can evolve from two different possible paths with the relative probability of $|a_1|^2:|a_2|^2$. The atoms in the final ensemble are, however, indistinguishable from one another and each is in a "mixed state" that is a superposition of two states Ψ_1 and Ψ_2, in the form of Eq. (2.4). The probability distribution function of the particles in the final state in the ensemble is, however, proportional to $|\Psi|^2$. It contains not only the terms $|a_1|^2|\Psi_1|^2 + |a_2|^2|\Psi_2|^2$ but also the cross terms, or the "interference terms" $(a_1^* a_2 |\Psi_1||\Psi_2|e^{-i(\phi_1 - \phi_2)} + a_1 a_2^* |\Psi_1||\Psi_2|e^{+i(\phi_1 - \phi_2)})$. Thus, $|\Psi|^2$ depends on, among other things, the relative phase $(\phi_1 - \phi_2)$ and the relative phases of a_1 and a_2. In short, since the probability distribution function is proportional to the square of the state function, whenever the final state function is a superposition of two or more state functions, the probability distribution function corresponding to the final state depends on the relative phases of the constituent state functions. It can lead to interference effects, much as in the familiar constructive and destructive interference phenomena involving electromagnetic waves. This is one of the manifestations of the wave–particle duality predicted by quantum mechanics and has been observed in numerous experiments. It has led to a great variety of important practical applications and is one of the major triumphs of quantum mechanics.

The superposition of states of two or more quantum systems that leads to correlated outcomes in the measurements of these systems is often described as "entanglement" in quantum information science in recent literature.

2.2 Operators (Postulate 2)

The second postulate states that *all physically "observable" properties of a dynamic system are represented by dynamic variables that are linear operators.* To understand what an operator is, let us look at what it does and what its mathematical properties and the corresponding physical interpretation are.

First, its connection to experimental results is the following. Consider, for example, an operator $\hat{Q}$ corresponding to the dynamic variable representing the property "Q" of the system. According to quantum mechanics, knowing the system is in the state Ψ does not mean that measurement of the property Q will necessarily yield a certain particular value. It will only tell us that repeated measurements of the same property Q of similar systems, or measurements of a large number of similar systems, all in the same state Ψ, will give a statistical distribution of values with an average value equal to:

$$\langle \hat{Q} \rangle_\Psi = \int \Psi^* \hat{Q} \Psi d\vec{r}, \tag{2.5a}$$

which is the expectation value of the property Q, and an uncertainty:

$$\Delta Q^2 = \int \Psi(\vec{r})^* (\hat{Q} - \langle \hat{Q} \rangle_\Psi)^2 \Psi(\vec{r}) d\vec{r}, \tag{2.5b}$$

in the three-dimensional Schrödinger representation. Because the operator $\hat{Q}$ is in general a function of the canonical variables $\hat{\vec{r}}$ and $\hat{\vec{p}}$ and can be a differential operator, one cannot arbitrarily reverse the order of multiplication of Ψ^* and $\hat{Q}$. Eqs. (2.5a & b) are generalizations of Eqs. (2.2) and (2.3) that introduced the concepts of expectation values and uncertainties.

Second, mathematically, an "operator" only has meaning when it operates on a state function. In general, an operator changes one state function to another. For example, the operator $\hat{Q}$ applied to an arbitrary state function Ψ generally changes it into another state Φ:

$$\hat{Q}\,\Psi = \Phi. \tag{2.6}$$

The true meaning of this simple abstract equation will not be clear until we know exactly how to find the operator expression representing each physical property. As corollaries of Postulate 2, there is a set of rules on how to do so for every dynamic property of the system.

Corollary 1

All the dynamic variables and, hence, all the corresponding operators representing any property of the system have the same functional dependence on the canonical variables *representing the position, $\hat{\vec{r}}$, and the linear momentum, $\hat{\vec{p}}$, as in classical mechanics. For example, the operators representing the kinetic energy, the angular momentum, the potential energy, and the Hamiltonian (total energy), etc. are, respectively,*

$$\hat{T} = \frac{\hat{P}^2}{2m}, \quad \hat{L} = \hat{\vec{r}} \times \hat{\vec{p}}, \quad \hat{V}(\hat{\vec{r}}), \quad and \quad \hat{H} = \hat{T} + \hat{V}, \ etc.$$

Corollary 2

In the Schrödinger representation, the operators representing the position coordinates $\hat{x}$, $\hat{y}$, $\hat{z}$, (or $\hat{\vec{r}}$) are the ordinary physical variables x, y, z, (or $\vec{r}$), but the operators representing the Cartesian components $\hat{p}_x$, $\hat{p}_y$, and $\hat{p}_z$ of the linear momentum, $\hat{\vec{p}}$, are the differential operators $(-i\hbar \frac{\partial}{\partial x})$, $(-i\hbar \frac{\partial}{\partial y})$, and $(-i\hbar \frac{\partial}{\partial z})$, respectively, (or $\hat{\vec{p}}$ is to be replaced by $-i\hbar\nabla$, whatever the coordinate system).

According to Newtonian mechanics, the dynamic properties of any system depend only on the position and the velocity (hence the position and the linear momentum) of the constituent parts of the system. Thus, on the basis of the above two corollaries of Postulate 2, the Schrödinger representation of the operator representing any dynamic property of the system is always known. With this knowledge, the innocent-looking simple "operator equation," (2.6), is now pregnant with profound physical meanings. For the operator $\hat{Q}$ is also to be interpreted physically as the process of measuring the property Q. Thus, the physical interpretation of the operator equation (2.6) is that, in the atomic and subatomic world, the process of measuring the property Q when the system is in the state Ψ generally changes it into another state Φ. Furthermore, once the Schrödinger representation of any operator is specified, Eq. (2.6) gives, mathematically, the exact effect the corresponding measurement process will have on the system in any particular state. It predicts that the measurement process $\hat{Q}$ will change the state of the system from Ψ to another state Φ, if the mathematical operation of $\hat{Q}$ on Ψ produces a function Φ that is not equal or proportional to Ψ. Two very important consequences follow from this consideration:

1. the notion of "commutation relationship" and the "uncertainty principle"; and
2. the concept of "eigen values and eigen functions."

Commutation relations and the uncertainty principle

An interesting question that can now be addressed is this. How does one know whether it is possible to have complete simultaneous knowledge of two specific properties of a system, say "A" and "B"?

Physically, for two properties to be specified simultaneously, it must be possible to measure one of the two properties without influencing the outcome of the measurement of the other property, and vice versa. In short, the order of measurements of the two properties should not matter, no matter what state the system is in. This means that application of the operator $\hat{A}\hat{B}$ on any arbitrary state Ψ should be exactly the same as applying the operator $\hat{B}\hat{A}$ on the same state, or:

$$\hat{A}\hat{B}\,\Psi = \hat{B}\hat{A}\,\Psi. \tag{2.7}$$

It in turn means that the effect of the operator $(\hat{A}\hat{B} - \hat{B}\hat{A})$ on any arbitrary state of the system must be equal to zero in this case:

$$[\hat{A}\hat{B} - \hat{B}\hat{A}]\Psi = 0.$$

The operator $(\hat{A}\hat{B} - \hat{B}\hat{A})$ itself is, therefore, equivalent to a "null operator":

$$[\hat{A}\hat{B} - \hat{B}\hat{A}] = 0 \tag{2.8}$$

when applied to any arbitrary state of the system, if the two properties can be specified precisely simultaneously.

The difference of two operators applied in different order is called the "commutator" of the operators $\hat{A}$ and $\hat{B}$ and defined as $[\hat{A}, \hat{B}]$:

$$\hat{A}\hat{B} - \hat{B}\hat{A} \equiv [\hat{A}, \hat{B}]. \tag{2.9}$$

When the commutator of two operators is equal to zero, the two operators are said to "commute." When two operators commute, as $\hat{A}$ and $\hat{B}$ in Eq. (2.8), it means that the two corresponding dynamic properties of the system can be measured in arbitrary order and specified precisely simultaneously, regardless of what state the system is in. There is now, therefore, a mathematically rigorous way to determine which two physical properties can be specified simultaneously and which ones may not be by simply calculating the commutator of the two corresponding operators.

In general, the commutator of two operators is not equal to zero but some third operator, say $\hat{C}$:

$$[\hat{A}, \hat{B}] = \hat{C}, \tag{2.10}$$

which can be evaluated mathematically on the basis of Postulate 2 and its corollaries. For example, let $\hat{A}$ be $\hat{x}$, and $\hat{B}$ be $\hat{p}_x$. From the Schrödinger representations of $\hat{x}$ and $\hat{p}_x$, it follows that:

$$(\hat{p}_x\hat{x})\Psi = -\,i\hbar\frac{\partial}{\partial x}[x\Psi(x)] = [-i\hbar x\frac{\partial}{\partial x}\Psi(x)] - i\hbar\Psi(x) = (x\hat{p}_x)\Psi - i\hbar\Psi;$$

therefore,

$$[\hat{x}, \hat{p}_x] = i\hbar. \tag{2.11a}$$

Similarly, one can derive the cyclic commutation relations among all the components of the position and momentum vectors:

$$[\hat{y}, \hat{p}_y] = i\hbar, \quad [\hat{z}, \hat{p}_z] = i\hbar,$$
$$[\hat{x}, \hat{y}] = 0, \quad [\hat{x}, \hat{z}] = 0, \quad [\hat{y}, \hat{z}] = 0,$$
$$[\hat{p}_x, \hat{p}_y] = 0, \quad [\hat{p}_x, \hat{p}_z] = 0, \quad [\hat{p}_y, \hat{p}_z] = 0. \qquad (2.11b)$$

Since all operators representing physically observable properties are functions of $\hat{x}$, $\hat{y}$, $\hat{z}$, $\hat{p}_x$, $\hat{p}_y$, and $\hat{p}_z$ only, one can obtain the commutator of any two operators on the basis of Postulate 2 or the commutation relationship (2.11a & 2.11b). Furthermore, it follows rigorously mathematically from the "Schwartz inequality" that, for any arbitrary state Ψ the system is in, the product of the uncertainties in any two operators as defined in (2.5b) is always equal to or greater than one half of the magnitude of the expectation value of the commutator:

$$(\Delta A)(\Delta B) \geq \frac{1}{2}|\langle[\hat{A}, \hat{B}]\rangle_\Psi| = \frac{1}{2}|\langle\hat{C}\rangle_\Psi|. \qquad (2.12)$$

(See, for example, Cohen-Tannoudji, *et al.* (1977), p. 287). Note that the "minimum uncertainty product" on the right side of Eq. (2.12) depends both on the commutator *and* the state the system is in. Thus, even if the two operators do not commute, their uncertainty product for a particular state can still be zero as long as *the expectation value of* the corresponding commutator in that *particular state* is equal to zero. In other states, the expectation value of the same commutator may not be zero. For commuting operators, even though the *minimum* uncertainty product is always zero, the uncertainty product *in general* may not be zero. Conversely, if the expectation of the corresponding commutator is not zero for all possible states, the *minimum* uncertainty product of two operators is never zero and, thus, the corresponding properties can never be specified precisely simultaneously. This is certainly the case, if the commutator of two non-commuting operators is a non-zero constant. For example, from Eqs. (2.11), the uncertainty product of $\hat{x}$ and $\hat{p}_x$ is always non-zero, and from (2.12):

$$\Delta x \Delta p_x \geq \hbar/2, \text{similarly for the } y \text{ and } z \text{ coordinates.} \qquad (2.13)$$

Equation (2.13) gives the astonishing result that it is not possible to know, or to specify, the position and the linear momentum in the same direction of the particle simultaneously. The more one knows about one of the two, the less one knows about the other. Equation (2.13) or its more general form (2.12) is a formal statement of the "Heisenberg uncertainty principle." It is important to note that *Eq. (2.12) shows explicitly the direct connection between the uncertainty principle as embodied in (2.13) and the commutation relationships (2.11a) and (2.11b) of the corresponding measurement processes.* It reflects, therefore, two different but entirely equivalent interpretations of the uncertainty principle that are often quoted alternatively in the literature. Equation (2.13) states that the product of the uncertainties in the results of measurements of the position and momentum of the particles in the ensemble must be greater than or equal to $\hbar/2$. Equation (2.12) shows that this uncertainty principle (2.13) is at the same time a consequence of the commutation relationships (2.11a) and (2.11b), which says that measurements of the position and the momentum of the particle are not independent of each other.

One might question whether the key result (2.13) of the uncertainty principle is physically reasonable. In the macroscopic world, if, for example, there is a billiard ball sitting there and not moving, one will certainly know it by simply looking at the ball. If it is in pitch darkness, one will not know either its position or its velocity. To know its position by looking, photons from some light source must be scattered from the billiard ball into the eye balls of the person doing the looking. Scattering photons from the billiard ball is not going to change its velocity. Because even though the photons have momentum, the mass of a macroscopic billiard ball is always too large for it to be moved any measurable amount by the momentum imparted to it by the photons. Thus, one can know its position and velocity simultaneously. Why is it then one cannot specify the position and velocity of an atomic or subatomic particle simultaneously?

A qualitative appreciation of the uncertainty principle might be gained on the basis of its interpretation based on the commutation relationships (2.11a) and (2.11b) of the corresponding measurement processes. Thus, consider, for example, instead of a billiard ball, a tiny atomic or subatomic particle. In the process of scattering at least one photon from the particle to a photodetector in order to measure its position, momentum will be transferred from the photon to the particle, the amount of which is not negligible for atomic or subatomic particles but is uncertain and depends upon the accuracy of the position measurement. (For a more in-depth discussion of this issue, see, for example, Bohm (1951). Section 5.11.) Subsequent measurement of the velocity of the particle will then give a result that is not the same as that when the position of the particle is determined. Thus, the position and the velocity of the atomic or subatomic particle cannot be specified precisely simultaneously. This example gives an intuitive basis for understanding *the uncertainty principle as embodied in Eq. (2.13) on the basis of its subtle connection, through Eq. (2.12), with the commutation properties of the operators representing the corresponding measurement processes.*

The basic commutation relationships, (2.11a & b), between the canonical variables, $\hat{x}$ and $\hat{p}_x$, and Heisenberg's uncertainty principle, (2.13), are both necessary consequences of the basic postulate that, in the Schrödinger representation, the operators $\hat{x}$ and $\hat{p}_x$ are x and $(-i\hbar\frac{\partial}{\partial x})$, respectively:

$$\hat{x} = x, \tag{2.14a}$$

$$\hat{p}_x = \left(-i\hbar\frac{\partial}{\partial x}\right), \text{ etc.} \tag{2.14b}$$

In fact, postulating any one of the three sets of the equations (2.11a, b), (2.13), or (2.14a, b), the other two will follow as consequences. Thus, any one of the three can be viewed as a part of the basic Postulate 2 of quantum mechanics. And, on the basis of

Postulate 2 in either form, it is possible to determine what physical properties can always be measured in arbitrary order and possibly be specified precisely simultaneously and which ones cannot.

Eigen values and eigen functions

With Postulates 1 and 2, another set of questions with great physical significance can be addressed. What properties of a system are quantized, what are not, and why? If a property is quantized, what possible results will measurements of such a property yield? These questions can now be answered precisely mathematically. *The allowed values of any property (or the result of any measurement of the property) are limited to the eigen values of the operator representing this property. If the corresponding eigen values are discrete, this property is quantized; otherwise, it is not.* What, then, are the "eigen values" and "eigen functions" of an operator? ("Eigen" came from the German word "Eigentum" that does not seem to have a precise English translation. It means something like "characteristic" or "distinct," or more precisely the "idio" part of "idiosyncrasy" in Greek, but its precise interpretation is probably best inferred from how it is used in context.)

As stated earlier, in general an operator operating on an arbitrary state function will change it to another state function. It can be shown that, associated with each operator representing a physically observable property, there is a unique set of characteristic state functions that will not change when operated upon by the operator. These state functions are called the "eigen functions" of this operator. Application of such an operator on each of its eigen functions leads to a characteristic number, which is a real number (no imaginary part), multiplying this eigen function. The characteristic number corresponding to each eigen function of the operator is called the "eigen value" corresponding to this eigen function. For example, the eigen value equation with discrete eigen values:

$$\hat{Q}\,\Psi_{q_i} = q_i\,\Psi_{q_i}, \qquad (i = 1, 2, 3, \ldots) \tag{2.15a}$$

gives the effect of an operator $\hat{Q}$ on its eigen function Ψ_{q_i} corresponding to the eigen value q_i. For continuous eigen values:

$$\hat{Q}\,\Psi_q = q\,\Psi_q, \tag{2.15b}$$

where q is a continuous variable. For some operators, some of the eigen values are discrete while the others are continuous. Accordingly, the problem of determining the allowed values of any property of the system is now reduced to that of solving the eigen value equation of either the form (2.15a) or (2.15b), as the case may be. Note that Ψ_{qi} can always be normalized:

$$\int \Psi_{q_i}^* \Psi_{q_i} d\vec{r} = 1, \tag{2.15c}$$

by definition. Other formal properties of the eigen functions will be discussed in detail in Section 2.4 below.

Suppose now the system is in the eigen state Ψ_{q_i}, the expectation value will always be q_i with zero uncertainty, as can be shown by substituting Ψ_{q_i} in (2.5a) and (2.5b):

$$\langle \hat{Q} \rangle_{\Psi_{q_i}} = \int \Psi_{q_i}^* \hat{Q} \Psi_{q_i} \, d\vec{r} = q_i, \tag{2.5c}$$

and the corresponding uncertainty is:

$$\Delta Q^2 = \int \Psi_{q_i}^* (\hat{Q} - \langle \hat{Q} \rangle_{\Psi_{q_i}})^2 \Psi_{q_i} \, d\vec{r} = 0. \tag{2.5d}$$

The physical interpretation of the eigen value equation is now clear. *If the system is in a particular eigen state of an operator, then any subsequent measurement of the corresponding property will always yield a value equal to the eigen value corresponding to that particular eigen state with no uncertainty.* For example, in the case of (2.5c), if the system is in the particular state Ψ_{q_3}, then measurement of the property Q will always give the particular value q_3 and the state of the system will always remain in Ψ_{q_3} after such a measurement, according to (2.15a or 2.15b). If the system is in an arbitrary unknown state Ψ, then any particular measurement of the property Q can yield any one of the quantized values q_i. In fact, to prepare a system in the particular state Ψ_{q_3} in the first place, one can simply choose the system in the state which measurement of the property Q yields the value q_3.

The discussion at this point may seem somewhat abstract, but the physical implications of all of this are profound and wide ranging. We will see many examples of eigen value equations for real properties of dynamic systems in later chapters. A fuller discussion of the mathematical properties of eigen values and eigen functions will be given in Section 2.4.

2.3 Equations of motion (Postulate 3)

The third postulate states that: *All state functions satisfy the "**time-dependent Schrödinger equation**":*

$$i\hbar \frac{\partial}{\partial t} \Psi = \hat{H} \Psi, \tag{2.16}$$

where $\hat{H}$ is the Hamiltonian of the system. The Hamiltonian is the operator corresponding to the total energy of the system. From Postulate 2, it is always possible to write down such an operator for any physical system of interest. Postulate 1 tells us that the state of any system is completely specified by the state function. Thus, solution of Schrödinger's equation for $\Psi(\vec{r}, t)$ describes completely the state of the dynamic system at all times once the initial condition $\Psi(\vec{r}, t = 0)$ and the Hamiltonian are known.

The Hamiltonian in general can be a function of time, $\hat{H}(\vec{r}, t)$. In the case when it is not a function of time (the system is "conservative" or the potential energy of the system $\hat{V}(\vec{r})$ is a function of the position coordinates only), the time-dependent Schrödinger equation is:

$$i\hbar \frac{\partial}{\partial t} \Psi(\vec{r}, t) = \left[-\frac{\hbar^2}{2m} \nabla^2 + V(\vec{r}) \right] \Psi(\vec{r}, t). \tag{2.17}$$

Equation (2.17) is of the form that can be solved by the method of separation of variables and the general solution is of the form:

$$\Psi(\vec{r}, t) = \sum_i C_i \Psi_{E_i}(\vec{r}, t) = \sum_i C_i \Phi_{E_i}(t) \Psi_{E_i}(\vec{r}). \tag{2.18}$$

$\Psi_{E_i}(\vec{r})$ is an eigen function of the Hamiltonian, or a solution of the "***time-independent Schrödinger equation***":

$$\hat{H}(\vec{r}) \Psi_{E_i}(\vec{r}) = E_i \Psi_{E_i}(\vec{r}), \tag{2.19}$$

and $\Phi_{E_i}(t)$ satisfies the equation:

$$i\hbar \frac{\partial \Phi_{E_i}(t)}{\partial t} = E_i \Phi_{E_i}(t). \tag{2.20}$$

Note that all state functions satisfy the time-dependent Schrödinger equation, (2.16); only the eigen functions of the Hamiltonian satisfy the time-independent Schrödinger equation, (2.19).

Equation (2.20) can be solved immediately to give:

$$\Phi_{E_i}(t) = e^{-\frac{i}{\hbar} E_i t},$$

for the initial condition $\Phi_{E_i}(0) = 1$. Thus, the time-dependence of the solution of the time-independent Schrödinger equation is simply:

$$\Psi_{E_i}(\vec{r}, t) = \Psi_{E_i}(\vec{r}) e^{-\frac{i}{\hbar} E_i t}. \tag{2.21}$$

Thus, if at $t = 0$, the system is in a particular eigen state of the Hamiltonian corresponding to the eigen value E_i:

$$\Psi(\vec{r}, t = 0) = \Psi_{E_i}(\vec{r}, t = 0),$$

the corresponding probability distribution is independent of time:

$$|\Psi_{E_i}(\vec{r}, t)|^2 = |\Psi_{E_i}(\vec{r})|^2$$

from (2.21). Thus, the eigen state of the Hamiltonian is called a "stationary state" of the system.

On the other hand, if the system is initially in a superposition state of the form (2.4), for example:

$$\Psi(\vec{r}, t=0) = a_m \Psi_{E_m}(\vec{r}) + a_n \Psi_{E_n}(\vec{r}),$$

at some time t later, the state function becomes:

$$\Psi(\vec{r}, t) = a_m \Psi_{E_m}(\vec{r}) e^{-\frac{i}{\hbar}E_m t} + a_n \Psi_{E_n}(\vec{r}) e^{-\frac{i}{\hbar}E_n t}. \tag{2.22}$$

The corresponding probability distribution function is:

$$|\Psi(\vec{r}, t)|^2 = |a_m \Psi_{E_m}(\vec{r})|^2 + |a_n \Psi_{E_n}(\vec{r})|^2 + a_m^* a_n \Psi_{E_m}^* \Psi_{E_n} e^{-\frac{i}{\hbar}(E_n - E_m)t}$$
$$+ a_m a_n^* \Psi_{E_m} \Psi_{E_n}^* e^{\frac{i}{\hbar}(E_n - E_m)t}. \tag{2.23}$$

It contains time-varying terms oscillating at the angular frequency, $\omega_{mn} = (E_m - E_n)/\hbar$, and is, therefore, "not stationary" in time.

The fact that the eigen states of the Hamiltonian are stationary states of the system has profound implications in understanding the structure and properties of all matters. Since the stable structure of any matter does not change with time, it must correspond to a stationary state and the lowest energy eigen state of the corresponding Hamiltonian. Thus, *the structures and properties of atoms, molecules, solids, or any other steady-state forms of matter can, in principle, be understood and explained on the basis of the solutions of the corresponding time-independent Schrödinger equations*, as we will see again and again in later chapters.

In summary, for an arbitrary initial state function $\Psi(\vec{r}, t=0)$ of a conservative system, the corresponding time-dependent Schrödinger equation can always be solved, if the initial state function can be expanded as a superposition of the eigen functions of the Hamiltonian:

$$\Psi(\vec{r}, 0) = \sum_n C_n \Psi_{E_n}(\vec{r}, 0).$$

In this case, $\Psi(\vec{r}, t)$ at an arbitrary time t is immediately known:

$$\Psi(\vec{r}, t) = \sum_n C_n \Psi_{E_n}(\vec{r}, t) = \sum_n C_n \Psi_{E_n}(\vec{r}) e^{-\frac{i}{\hbar}E_n t} \tag{2.24}$$

from (2.21), where C_n is a constant independent of time. The validity of this solution can be confirmed by substituting (2.24) into (2.16) and making use of (2.19). *Solution of the time-independent Schrödinger equation is, thus, the first step in solving the time-dependent Schrödinger equation.*

It turns out that it is always possible, in principle, to expand any state function in terms of the eigen functions of the Hamiltonian in the form of Eq. (2.24), as will be shown in Section 2.4 below; therefore, *it is always possible to solve the time-dependent Schrödinger equation in the form of Eq.* (2.24) provided the time-independent Schrödinger equation can be solved. To show how that is done in more detail, we must first discuss the mathematical properties of the eigen functions.

2.4   Eigen functions, basis states, and representations

This section is devoted to some of the mathematical properties of eigen functions and the related expansion theorem. These are the basic tools for solving time-dependent and time-independent Schrödinger equations and, thus, many of the important problems in the applications of quantum mechanics, as will be shown in later chapters.

As the above discussion in connection with the solution of the time-dependent Schrödinger equation showed, the key to its solution is that it must be possible to expand an arbitrary state function in terms of the eigen functions of the Hamiltonian. The reason that this is always possible is that the eigen functions of not only the Hamiltonian but all the operators corresponding to physical observables form a "*complete orthonormal set.*" It means that the eigen functions:

1. are "orthogonal" to each other,
2. can always be "normalized," and
3. form "a complete set."

While a rigorous mathematical proof of this statement is not of particular interest here, it is important to see what each of these properties mean precisely and how they are used in solving problems.

For definiteness, let us start from Eq. (2.15a) for the discrete eigen value case:

$$\hat{Q}(\vec{r})\Psi_{q_i}(\vec{r}) = q_i \Psi_{q_i}(\vec{r}).$$

"Orthonormality" means that the eigen functions have the property:

$$\int \Psi_{q_i}^*(\vec{r})\Psi_{q_j}(\vec{r})\mathrm{d}\vec{r} = \delta_{ij} \equiv \begin{cases} 1, & if \ i = j \\ 0, & if \ i \neq j \end{cases}. \tag{2.25}$$

Completeness means that the "delta function," which is the 'sharpest possible function of unit area,' can be constructed from the complete set of eigen functions:

$$\sum_n \Psi_{q_n}^*(\vec{r})\Psi_{q_n}(\vec{r}') = \delta(\vec{r} - \vec{r}') = \delta(x - x')\delta(y - y')\delta(z - z'). \tag{2.26}$$

$\delta(\vec{r})$ here represents the Dirac delta function, which means:

$$\Psi(\vec{r}) = \int \Psi(\vec{r}')\delta(\vec{r}' - \vec{r})\mathrm{d}\vec{r}' \tag{2.27}$$

for any arbitrary state function $\Psi(\vec{r})$. By multiplying the right and left sides of (2.25) by $\Psi_{q_i}(\vec{r}')$ followed by summing over q_i, it can be shown that the completeness relation (2.26) follows from the orthogonality condition (2.25).

Substituting (2.26) into (2.27) gives:

$$\Psi(\vec{r}) = \sum_n C_n \Psi_{q_n}(\vec{r}), \tag{2.28}$$

where

$$C_n = \int \Psi^*_{q_n}(\vec{r})\Psi(\vec{r})d\vec{r}. \tag{2.29}$$

Thus, any arbitrary state function can always be expanded as a superposition of a *complete* set of orthonormal eigen functions. This is another way of saying that the set of eigen functions is "complete," in the sense that *any arbitrary function* can always be constructed as a superposition of such a set of functions. The eigen functions in such an expansion are known as the "basis states." Since knowing the expansion coefficients is tantamount to knowing the state function itself, the set of expansion coefficients is a "representation" of the state function using this particular basis of expansion. For example, if the operator $\hat{Q}$ is the Hamiltonian $\hat{H}$ of the system in the example above, the corresponding expansion (2.28) is then:

$$\Psi(\vec{r}) = \sum_n C_n \Psi_{E_n}(\vec{r}), \tag{2.28a}$$

where

$$C_n = \int \Psi^*_{E_n}(\vec{r})\Psi(\vec{r})d\vec{r}. \tag{2.29a}$$

In (2.28a) the set of C_n then gives the state function Ψ in the "energy representation," and the Ψ_{E_n} are the "basis states" in the energy representation. The C_n as given by the scalar product shown in (2.29a) can be viewed as the projections of the state function Ψ along the 'coordinate axis' represented by the eigen functions Ψ_{E_n} in a multidimensional "Hilbert space." If the number of eigen values is N, then it is an N-dimensional Hilbert space. The expansion coefficient C_n has a very simple physical interpretation. Its magnitude squared $|C_n|^2$ is the relative probability that measurement of the energy of the system will yield the value E_n. The absolute probability is then $|C_n|^2/\sum_i |C_i|^2$. If the state function is normalized, or $\sum_i |C_i|^2 = 1$, then the value of $|C_n|^2$ is the absolute probability. Similarly, in the general case where the basis states are the eigen functions of the operator $\hat{Q}$ as in (2.28), the square of the expansion coefficient C_i gives the probability that measurement of the property Q gives the value q_i.

In the case when the eigen value is continuous, the sums in Eqs. (2.28) and (2.26) are replaced by integrals. For example, in the case of Eq. (2.15b):

$$\hat{Q}\Psi_q(\vec{r}) = q\Psi_q(\vec{r}),$$

the Q-representation of the state function would be $C(q)$:

$$C(q) = \int \Psi^*_q(\vec{r})\Psi(\vec{r})\,d\vec{r}, \tag{2.28b}$$

and

$$\Psi(\vec{r}) = \int C(q)\Psi_q(\vec{r})\mathrm{d}q. \tag{2.29b}$$

In analogy with the discrete eigen value case, the physical interpretation of $|C(q)|^2$ is that it is the probability that measurement of the property Q of the particle in the normalized state Ψ will yield the value between q and $q + \mathrm{d}q$. The corresponding completeness condition is:

$$\int \Psi_q^*(\vec{r})\Psi_q(\vec{r}')\mathrm{d}q = \delta(\vec{r} - \vec{r}'), \tag{2.30}$$

and the orthonormality condition becomes:

$$\int \Psi_q^*(\vec{r})\Psi_{q'}(\vec{r})\mathrm{d}\vec{r} = \delta(q - q'). \tag{2.31}$$

The concept of "representation" of a state function in quantum mechanics is very much like, for example, the concept of representing a time-dependent signal by a Fourier series or integral in electrical engineering. Knowing the spectrum of the signal in the frequency domain is tantamount to knowing the time-dependent signal itself. In quantum mechanics, the wave functions can be represented by the coefficients of expansion in different representations. The fact that the same state can have different representations plays a key role in the recently proposed scheme of quantum cryptography.

2.5 Alternative notations and formulations

The basic postulates and rules of algebra for quantum mechanics have so far all been given in terms of state functions and operators in the Schrödinger representation, because the vast majority of the practical problems in solid state electronics and photonics can all be adequately dealt with using this formulation. There are, however, problems that can be handled more conveniently using alternative, but completely equivalent formulations, such as Heisenberg's formulation of quantum mechanics using matrices, which is sometimes known as matrix mechanics. In terms of notations also, as the problems become more complicated, as practical problems always will be, there is a real need to simplify and eliminate superfluous information from the notations. The Dirac notation is an elegant system of compact notations that is widely used in quantum mechanics, without which written equations in quantum theory will become impossibly unwieldy, as we will see later. We will introduce this efficient system first and then consider briefly Heisenberg's matrix formulation of quantum mechanics.

Dirac's notation

The Dirac notation of the abstract state function Ψ is either a "bra" vector $\langle\Psi|$ or a "ket" vector $|\Psi\rangle$ regardless of what representation it is in. The distinction between the two forms lies in how the state vector is used and will become clear when they are used again and again in different contexts.

The scalar product of two state functions in the Schrödinger representation Φ and Ψ:

$$\int \Phi^*(\vec{r})\Psi(\vec{r})d\vec{r}$$

in the Dirac notation is the "bracket" $\langle\Phi|\Psi\rangle$, which is the scalar product of the bra vector $\langle\Phi|$ and the ket vector $|\Psi\rangle$. The bracket is by definition the corresponding integral and it is an ordinary number. As far as the final numerical result of the integral is concerned, the information on what coordinate system is used in carrying out the integration is superfluous, which may be, for example, the Cartesian, or cylindrical, or spherical system. In short, since the choice is not unique, in the Dirac notation it is suppressed and by definition:

$$\int \Phi^*(\vec{r})\Psi(\vec{r})d\vec{r} \equiv \langle\Phi|\Psi\rangle. \tag{2.32}$$

Suppose the state function Ψ in (2.32) is generated from another state function Ψ' by an operator $\hat{Q}$ as given by the operator equation of the form (2.6):

$$\hat{Q}\Psi' = \Psi.$$

The scalar product in (2.32) then becomes:

$$\int \Phi^*(\vec{r})\Psi(\vec{r})d\vec{r} = \int \Phi^*(\vec{r})\hat{Q}\Psi'(\vec{r})d\vec{r}.$$

In the Dirac notation, the integral above is, by definition:

$$\int \Phi^*(\vec{r})\hat{Q}\Psi'(\vec{r})d\vec{r} \equiv \langle\Phi|\hat{Q}|\Psi'\rangle. \tag{2.33}$$

Again, the notation is more compact and any superfluous information is not carried along.

The time-independent Schrödinger equation in the Dirac notation, for example, is:

$$\hat{H}|E_i\rangle = E_i|E_i\rangle, \tag{2.34}$$

in which the eigen function is indicated only by its corresponding eigen value and the Dirac notation for the state function. As another example, the eigen value equation for the operator $\hat{x}$ with its continuous eigen value x and the corresponding eigen function $|x\rangle$ is:

$$\hat{x}|x\rangle = x|x\rangle. \tag{2.35}$$

The state function $\Psi(x)$ is the projection of $|\Psi\rangle$ on the eigen function $|x\rangle$; therefore,

$\Psi(x)$ *in the Dirac notation is* $\langle x|\Psi\rangle$, (2.36a)

and its complex conjugate $\Psi^*(x)$ *is* $\langle \Psi|x\rangle$. (2.36b)

The orthonormality condition for the case of discrete eigen values is, for example:

$$\langle E_i|E_j\rangle = \delta_{ij},$$ (2.37)

and for the continuous eigen value case is, for example:

$$\langle x|x'\rangle = \delta(x - x').$$ (2.38)

The completeness condition (2.26) in the one-dimensional case becomes:

$$\sum_n \langle x|E_n\rangle\langle E_n|x'\rangle = \delta(x - x'),$$ (2.39)

which can also be written as

$$\langle x|\left(\sum_n |E_n\rangle\langle E_n|\right)|x'\rangle = \delta(x - x').$$ (2.40)

Comparison of (2.40) and (2.38) shows that the factor in the parentheses on the left side of (2.40) must have the meaning of a "unit operator":

$$\sum_n |E_n\rangle\langle E_n| = \hat{1}.$$ (2.41)

The analogous result in the case of an operator such as $\hat{x}$ with a continuous eigen value x is:

$$\int |x\rangle\langle x|\mathrm{d}x = \hat{1}.$$ (2.42)

These alternative expressions of the completeness condition are extremely useful tools for arriving at the expansion of state functions in different bases such as the form (2.28) and (2.29) or (2.28a) and (2.29a). For example, applying the unit operators (2.41) and (2.42) to an arbitrary state vector $|\Psi\rangle$ gives immediately the expansion of this state vector in the representation with the eigen states of the Hamiltonian as the basis:

$$\Psi(x) = \langle x|\Psi\rangle = \langle x|\ \hat{1}\ |\Psi\rangle = \sum_n \langle x|E_n\rangle\langle E_n|\Psi\rangle$$
$$= \sum_n C_n\langle x|E_n\rangle = \sum_n C_n\Psi_{E_n}(x),$$

where

$$C_n = \langle E_n|\Psi\rangle = \langle E_n|\hat{1}|\Psi\rangle = \langle E_n \,|(\int |x\rangle\langle x|\mathrm{d}x) \,|\Psi\rangle$$

$$= \int (\,\langle E_n|x\rangle\langle x|\Psi\rangle\,)\,\mathrm{d}x = \int \Psi^*_{E_n}(x)\Psi(x)\mathrm{d}x,$$

which are exactly the same as in (2.28a) and (2.29a). Similarly, *expressing the unit operator in any representation with discrete or continuous eigen values of the form* (2.41) and (2.42), respectively, can lead immediately to the expansion theorem in any representation.

With these powerful tools, we can now introduce the basic concepts of Heisenberg's formulation of quantum mechanics in terms of matrices. There are no new postulates, only the mathematics is in different but equivalent forms.

Heisenberg's matrix formulation of quantum mechanics

The key point is that *the state function* $|\Psi\rangle$ can be represented as a vector by its projections on a complete set of basis states, for example, the eigen functions of the Hamiltonian, $C_n = \langle E_n|\Psi\rangle$, or some other operator of choice. This means that the ket vector $|\Psi\rangle$ is a vector in matrix algebra:

$$|\Psi\rangle = \begin{pmatrix} \langle E_1|\Psi\rangle \\ \langle E_2|\Psi\rangle \\ \cdots \\ \cdots \end{pmatrix}.$$

In the same representation, the basic operator equation (2.6):

$$\hat{Q}\,\Psi = \Phi \tag{2.6}$$

becomes a matrix equation, which can be derived from (2.6) by multiplying from the left by the bra vector $\langle E_n|$:

$$\langle E_n|\hat{Q}|\Psi\rangle = \langle E_n|\Phi\rangle.$$

Inserting a unit operator in the above equation:

$$\langle E_n|\hat{Q}\cdot\hat{1}|\Psi\rangle = \langle E_n|\Phi\rangle$$

in the same representation, and making use of the completeness condition, (2.41), we have:

$$\sum_m \langle E_n|\hat{Q}|E_m\rangle\langle E_m|\Psi\rangle = \langle E_n|\Phi\rangle, \tag{2.43}$$

which is a matrix equation:

$$
\begin{pmatrix}
\langle E_1|\hat{Q}|E_1\rangle & \langle E_1|\hat{Q}|E_2\rangle & \langle E_1|\hat{Q}|E_3\rangle & \cdots \\
\langle E_2|\hat{Q}|E_1\rangle & \langle E_2|\hat{Q}|E_2\rangle & \cdots & \cdots \\
\cdots & \cdots & \cdots & \cdots \\
\cdots & \cdots & \cdots & \cdots
\end{pmatrix}
\begin{pmatrix}
\langle E_1|\Psi\rangle \\
\langle E_2|\Psi\rangle \\
\cdots \\
\cdots
\end{pmatrix}
$$

$$
=
\begin{pmatrix}
\langle E_1|\Phi\rangle \\
\langle E_2|\Phi\rangle \\
\cdots \\
\cdots
\end{pmatrix} .
$$

(2.43a)

Thus, in this representation, the state functions and operators have all become column vectors and square matrices, respectively. Of particular interest is the matrix representation of $\hat{H}$ in this representation. Because the basis states $|E_n\rangle$ are the eigen states of $\hat{H}$, its matrix representation is "diagonal":

$$
\begin{pmatrix}
E_1 & 0 & 0 & 0 & \cdots\cdots \\
0 & E_2 & 0 & 0 & \cdots\cdots \\
0 & 0 & E_3 & 0 & \cdots\cdots \\
\cdots & \cdots & \cdots & \cdots & \cdots \\
\cdots & \cdots & \cdots & \cdots & \cdots
\end{pmatrix} ,
$$

and the diagonal elements are the eigen values of $\hat{H}$. The terminology is that this is "the representation in which $\hat{H}$ is diagonal." In the matrix formulation, "diagonalizing the matrix" according to the rules of matrix algebra is, therefore, equivalent to solving the time-independent Schrödinger equation.

In the representation in which $\hat{H}$ is diagonal, the scalar product $\langle\Phi|\Psi\rangle$ is:

$$
\langle\Phi|\Psi\rangle = \langle\Phi|\hat{1}|\Psi\rangle = \sum_n \langle\Phi|E_n\rangle\langle E_n|\Psi\rangle
$$

$$
= (\langle\Phi|E_1\rangle\langle\Phi|E_2\rangle \cdots)
\begin{pmatrix}
\langle E_1|\Psi\rangle \\
\langle E_2|\Psi\rangle \\
\cdots \\
\cdots
\end{pmatrix} ,
$$

(2.44)

which is the scalar product of a row vector and a column vector in matrix algebra. Thus, in the matrix representation, the bra vector $\langle\Phi|$ is a row vector, while the ket vector $|\Psi\rangle$ is a column vector. It distinguishes one from the other, even though both represent state functions.

Note also the absolute importance of the order of multiplication of the bra vectors and ket vectors. Product of a bra vector on the left with a ket vector on the right results in a bracket which is a scalar or a number, as in (2.32) and (2.44). Product of a ket vector on the left and a bra vector on the right leads to a matrix or an operator, as in (2.41) and (2.42).

With the matrix representations of state functions and operators in the form of Eq. (2.43a), the related operations in quantum mechanics described in the previous sections can all be carried out and expressed according to the rules of matrix algebra, in complete equivalence to those in the Schrödinger representation. This is the basis of Heisenberg's formulation of quantum mechanics using matrices. It is a widely used approach to solving practical problems, especially in dealing with, for example, problems involving the angular momentum of atoms and molecules or the interaction of electromagnetic radiation and matters of different forms, as we shall see later.

Heisenberg's equation of motion

Consider now the equations of motion. In Schrödinger's picture, the dynamics of the system is completely described by the time-dependent Schrödinger equation for the wave function. In the alternative Heisenberg's picture, it is described by "Heisenberg's equation of motion for the dynamic variables." The link between the two is that both approaches ultimately give the same numerical results for the expectation values of any observable of the system at all times. Therefore, insofar as experimental results are concerned, it makes no difference which equation of motion is used to describe the results.

Consider, for example, an arbitrary property represented by the operator $\hat{Q}$, which does not have any explicit time-dependence. Suppose at $t = 0$ the system is in the state $|\Psi_0\rangle = \Psi(\vec{r}, 0)$ and the expectation value of $\hat{Q}$ is $\langle\Psi_0|\hat{Q}|\Psi_0\rangle$. In Schrödinger's picture, the state function is a function of time, $|\Psi_t\rangle \equiv \Psi(\vec{r}, t)$. The only time variation in the expectation value comes from the time dependence in the state function $|\Psi_t\rangle$. In contrast, in Heisenberg's picture, the state function does not vary with time and remains $|\Psi_0\rangle$, the same time variation in the expectation value due to the time variation in the state function in the Schrödinger picture is now ascribed to a time-dependent dynamic variable $\hat{Q}_t$, which at $t = 0$ is the same as the operator $\hat{Q}$ in Schrödinger's picture. If the two pictures are such that they give exactly the same expectation value at all times:

$$\langle\Psi_0|\hat{Q}_t|\Psi_0\rangle \equiv \langle\Psi_t|\hat{Q}|\Psi_t\rangle, \tag{2.45}$$

they must be describing the same dynamics. Let us now express the above equation in the representation in which $\hat{H}$ is diagonal:

$$\sum_{nn'} \langle\Psi_0|E_n\rangle\langle E_n|\hat{Q}_t|E_{n'}\rangle\langle E_{n'}|\Psi_0\rangle$$
$$\equiv \sum_{mm'} \langle\Psi_t|E_m\rangle\langle E_m|\hat{Q}|E_{m'}\rangle\langle E_{m'}|\Psi_t\rangle. \tag{2.46}$$

From solutions of the time-dependent Schrödinger equation (2.21), we know that:

$$\langle\Psi_t|E_{m'}\rangle = e^{\frac{i}{\hbar}E_{m'}t}\langle\Psi_0|E_{m'}\rangle \tag{2.47a}$$

and

$$\langle E_{m'}|\Psi_t\rangle = e^{-\frac{i}{\hbar}E_{m'}t}\langle E_{m'}|\Psi_0\rangle.$$ (2.47b)

Substituting (2.47a) and (2.47b) for all the state functions into (2.46) gives:

$$\sum_{nn'} \langle\Psi_0|E_n\rangle \langle E_n|\hat{Q}_t|E_{n'}\rangle \langle E_{n'}|\Psi_0\rangle$$
$$\equiv \sum_{mm'} \langle\Psi_0|E_m\rangle e^{\frac{i}{\hbar}E_m t}\langle E_m|\hat{Q}|E_{m'}\rangle e^{-\frac{i}{\hbar}E_{m'}t}\langle E_{m'}|\Psi_0\rangle.$$ (2.48)

Since $|\Psi_0\rangle$ is totally arbitrary, similar terms in the sums on both sides of (2.48) must be equal to each other term-by-term and, therefore,

$$\langle E_n|\hat{Q}_t|E_{n'}\rangle = e^{\frac{i}{\hbar}E_n t} \langle E_n|\hat{Q}|E_{n'}\rangle e^{-\frac{i}{\hbar}E_{n'}t}.$$

Differentiating the above equation and making use of the time-independent Schrödinger equation gives:

$$\frac{\mathrm{d}}{\mathrm{d}t}\langle E_n|\hat{Q}_t|E_{n'}\rangle = \frac{i}{\hbar}(E_n - E_{n'})\langle E_n|\hat{Q}_t|E_{n'}\rangle$$
$$= \frac{i}{\hbar}\langle E_n|(\hat{H}\hat{Q}_t - \hat{Q}_t\hat{H}|E_{n'}\rangle,$$

which leads to the operator equation:

$$\frac{\mathrm{d}}{\mathrm{d}t}\hat{Q}_t = \frac{i}{\hbar}(\hat{H}\hat{Q}_t - \hat{Q}_t\hat{H}) \equiv \frac{i}{\hbar}[\hat{H},\hat{Q}_t].$$ (2.49)

One important result immediately follows from this equation of motion: *Any operator that commutes with the Hamiltonian represents a physical property of the system that is a constant of motion, or $\frac{\mathrm{d}}{\mathrm{d}t}\hat{Q}_t = 0$,* just like the total energy of the system which is represented by the Hamiltonian itself. Since the eigen values of the complete set of commuting operators including the Hamiltonian are needed to characterize completely the state of a dynamic system, the states of the system are specified by the eigen values of all the constants of motion of the system. The corresponding eigen values of the complete set of commuting operators are sometimes referred to as the "good quantum numbers" of the state.

If $\hat{Q}$ has an explicit time-dependence because of some external parameter not directly dependent on the dynamics of the system under the influence of the action represented by the Hamiltonian $\hat{H}$, then the above equation must be modified to take that into account. It results in the equation of motion in Heisenberg's picture of the form:

$$\frac{\mathrm{d}}{\mathrm{d}t}\hat{Q}_t = \frac{i}{\hbar}[\hat{H},\hat{Q}_t] + \left(\frac{\partial}{\partial t}\hat{Q}_t\right),$$ (2.50)

which is also known as "Heisenberg's equation of motion." Thus, the scheme using the matrix representations of the state functions, operators, and Heisenberg's equation of motion is a complete alternative approach to that of Schrödinger's formulation for solving all quantum mechanic problems.

Note that, in Heisenberg's picture, the dynamics of the system is described by the time dependence of the dynamic variables, not the state functions. In that sense, it superficially resembles the picture in classical mechanics in which the dynamics is described by the time dependence of the dynamic variables x and v_x. There are, however, important differences in that the quantum mechanical dynamical variables, such as $\hat{Q}_t$, are operators or matrices, not ordinary physical variables. Also, Heisenberg's equation of motion (2.50) involves the Hamiltonian and the commutator and it is of a very different form from Newton's equation of motion for x or v_x, which involves the force. On the other hand, Heisenberg's equation is of the form of Hamilton's equation of motion in classical mechanics with the "Poisson bracket" playing the role of the commutator. We will not, however, digress into a discussion of this topic here; it is not germane to our main concerns and is outside the scope of this book.

Concluding remarks

The fact that there are different formulations of quantum mechanics is a two-edged sword. At first sight, it seems confusing that something as definitive as the laws of physics can be interpreted and treated mathematically this way or that way. Why is it necessary, or desirable, even to mention any alternative approach? The reward for this apparent confusion is that it offers great flexibility in treating different problems, as we will see in later chapters. One should not forget that, in classical mechanics also, Newton's formulation is not the only way to deal with mechanics problems in the macroscopic world. To be sure, it is the most convenient and useful one. It is the approach that is customarily taught in the schools from the earliest days on and is used to solve the simplest to the most challenging and complex problems in mechanics. There are, however, also Hamilton's and Lagrange's formulations of classical mechanics, which are occasionally used for specialized problems. In the case of quantum mechanics, however, both Schrödinger's formulation based on the wave functions and Heisenberg's matrix formulation complement each other and both are widely used.

Although a great deal more can be said about the basic postulates and the formalisms of quantum mechanics, the foundations for formulating and solving problems in the atomic and subatomic world have now been established. We will now proceed to use these basic principles to solve the problems of interest in the following chapters. We will see many unusual predictions about the world on the atomic and subatomic scale that are unfathomable from the point of view of classical mechanics, but are logical consequences of the basic postulates of quantum mechanics. Without exception, these all agree with experimental observations. There is no question that the principles of quantum mechanics describe the laws of Nature of the atomic and subatomic world.

2.6 Problems

2.1. Consider the following hypothetic wave function for a particle confined in the region $-4 \leq x \leq 6$:

$$\Psi(x) = \begin{cases} A(4+x), & -4 \leq x \leq 1 \\ A(6-x), & 1 \leq x \leq 6 \\ 0, & \text{otherwise.} \end{cases}$$

(a) Sketch the wave function.
(b) Normalize this wave function over the range the particle is confined in.
(c) Determine the expectation values $\langle x \rangle$, $\langle x^2 \rangle$, and $\sigma^2 \equiv \langle (x - \langle x \rangle)^2 \rangle$ using the normalized wave function.
(d) Again, using the normalized wave function, calculate the expectation value of the kinetic energy of the particle.

2.2. Suppose an electron is *confined* to a zero-potential region between two impenetrable walls at $x = 0$ and a for all times. Its initial wave function is given by:

$$\Psi(x) = \begin{cases} \sqrt{\frac{2}{a}} \sin(3\pi x/a), & \text{for} \quad 0 \leq x \leq a \\ 0, & \text{for} \quad x \leq 0 \text{ and } x \geq a. \end{cases}$$

(a) Calculate the expectation value of the energy $\langle \hat{H} \rangle$ in this state.
(b) Calculate the energy eigen value corresponding to this state.
(c) What is the time dependence of $\Psi(x, t)$?
(d) Calculate the uncertainty in the total energy in this state, $\Delta \hat{H}$. Is the answer what you expect? Explain.

2.3. Prove the following commutation relationships:

(a) $[\hat{A} + \hat{B}, \hat{C}] = [\hat{A}, \hat{C}] + [\hat{B}, \hat{C}]$,
(b) $[\hat{A}, \hat{B}\hat{C}] = [\hat{A}, \hat{B}]\hat{C} + \hat{B}[\hat{A}, \hat{C}]$.

2.4. Prove the following commutation relations:

(a) $[\hat{p}_x, \hat{x}^n] = -i\hbar n \hat{x}^{n-1}$.
(b) $[\hat{x}, \hat{p}_x^2] = i 2\hbar \hat{p}_x$.
(c) From (b) above, what can you say about the possibility of measuring the position and kinetic energy of a particle in an arbitrary state simultaneously with zero uncertainty in each measurement?

2.5. Consider the two-dimensional matrices $\hat{\sigma}_x = \begin{pmatrix} 0 & 1 \\ 1 & 0 \end{pmatrix}$, $\hat{\sigma}_y = \begin{pmatrix} 0 & -i \\ i & 0 \end{pmatrix}$, and $\hat{\sigma}_z = \begin{pmatrix} 1 & 0 \\ 0 & -1 \end{pmatrix}$, whose physical significance will be discussed later in Chapter 6

(a) Find the eigen values and the corresponding normalized eigen functions of these matrices in the matrix notation.
(b) Write these eigen functions in the Dirac notation in the representation in which $\hat{\sigma}_z$ is diagonal.

2.6. Consider the Hamiltonian operator $\hat{H}$ with discrete eigen values. Suppose we know that the Hamiltonian is a Hermitian operator which by definition satisfies the condition:

$$\int \Psi^*(x)\, \hat{H}\, \Phi(x)\mathrm{d}x = \left(\int \Phi^*(x)\, \hat{H}\, \Psi(x)\mathrm{d}x \right)^* .$$

Show that:

(a) The eigen values of the Hamiltonian are all real.
(b) The eigen functions of $\hat{H}$ corresponding to different eigen values must be orthogonal to each other.

2.7. Consider a particle of mass m in a potential field $V(x)$.

(a) Show that the time variation of the expectation value of the position is given by:

$$\frac{\mathrm{d}}{\mathrm{d}t}\langle x \rangle = \frac{\langle p_x \rangle}{m} .$$

(b) Prove that the time variation of the expectation value of the momentum is given by:

$$\frac{\mathrm{d}}{\mathrm{d}t}\langle p_x \rangle = - \left\langle \frac{\mathrm{d}V}{\mathrm{d}x} \right\rangle = F_x,$$

which is known as Ehrenfest's theorem.

3 Wave/particle duality and de Broglie waves

While the basic principles and mathematical tools of quantum mechanics are outlined in the previous chapter, it remains to be seen what the physical consequences are and how it deals with specific problems. It is shown in this chapter that a particle in motion in free space has wave properties. This wave/particle duality is a consequence of Heisenberg's uncertainty principle. Because particles are also waves, localized particles must be "wave packets" corresponding to superpositions of de Broglie waves, and the spatial Fourier transform of the wave function in real space is its momentum representation in the de Broglie wave-vector $\vec{k}$-space.

3.1 Free particles and de Broglie waves

One simple question that can be asked is what is the state function for a particle of mass m moving in free space with a constant velocity v_x or linear momentum $p_x = mv_x$. Based on the discussion in connection with Eqs. (2.15a) and (2.15b), this state function must be the eigen function corresponding to the eigen value p_x, which can be of a positive or negative value, of the operator $\hat{p}_x$ representing the x component of the linear momentum:

$$\hat{p}_x \Psi_{p_x}(x) = p_x \Psi_{p_x}(x). \tag{3.1}$$

Following Corollary 2 of Postulate 2 or as a necessary consequence of the commutation relationship (2.11a), (3.1) becomes a simple ordinary differential equation in the Schrödinger representation:

$$-i\hbar \frac{\partial}{\partial x} \Psi_{p_x}(x) = p_x \Psi_{p_x}(x) \tag{3.2}$$

with the solution

$$\Psi_{p_x}(x) = C e^{\frac{i}{\hbar} p_x x}, \tag{3.3}$$

where C is a constant to be determined by the normalization condition. Since the particle is in free space, there is no boundary condition on the state function; the velocity or linear momentum of the particle can have any value and is a continuous eigen value of the operator $\hat{p}_x$.

This simple result, (3.3), has some very interesting implications. First, the corresponding probability distribution function:

$$\left|\Psi_{p_x}(x)\right|^2 = |C|^2$$

is a constant independent of x, meaning the particle could equally well be anywhere. This is exactly what Heisenberg's uncertainty principle would have told us: if the linear momentum of the particle is known precisely to be of the value p_x, or the uncertainty $\Delta p_x = 0$, the uncertainty in the position, Δx, must be infinite and the particle can be anywhere with equal probability. This is what quantum mechanics says. How can one understand it intuitively? Superficially, it is not as unreasonable as one might think on first sight. As we recall, in quantum mechanics, the result of measurements must always be understood in the probability sense. "A particle traveling with a constant velocity v_x" always means "an ensemble of similar particles all in the same state of having the same linear momentum $p_x = mv_x$." Clearly, if this is the only information about the state of the particle, the particle in the ensemble can be anywhere with equal probability. It must be emphasized, however, that Heisenberg's uncertainty principle goes far deeper than this intuitive argument about semantics. The uncertainty principle states that it is fundamentally not even possible to specify the position and velocity precisely simultaneously.

The constant C in (3.3) can be fixed by the normalization condition. Since the distribution function is independent of x, if there is only one particle in all space from $x = -\infty$ to $+\infty$, $|C|^2$ must be zero. A more meaningful situation is when there are N particles per unit length in space or one particle in the space range $1/N$:

$$1 = \int_0^{1/N} \left|\Psi_{p_x}(x)\right|^2 \mathrm{d}x = |C|^2/N; \tag{3.4}$$

therefore, $|C|^2 = N$, or $C = (N)^{1/2}$. By convention, the phase of C can be chosen to be zero, since the expectation value of any observable always involves Ψ and Ψ^* in pairs and the reference phase factor cancels out. The state function:

$$\Psi_{p_x}(x) = \sqrt{N}\mathrm{e}^{\frac{i}{\hbar}p_x x} \tag{3.5}$$

describes, therefore, a beam of N particles per unit length traveling with the constant velocity $v_x = p_x/m$.

For a free particle, the Hamiltonian is:

$$\hat{H} = \frac{\hat{p}_x^2}{2m} = -\frac{\hbar^2}{2m}\frac{\partial^2}{\partial x^2}, \tag{3.6}$$

and the corresponding time-independent Schrödinger equation is:

$$\hat{H}\,\Psi_E(x) = \left[-\frac{\hbar^2}{2m}\frac{\partial^2}{\partial x^2}\right]\Psi_E(x) = E\,\Psi_E(x). \tag{3.7}$$

Substituting the state function (3.5) into (3.7) shows that it is also an eigen function of the Hamiltonian:

$$\Psi_E(x) = \sqrt{N}e^{\frac{i}{\hbar}p_x x}, \tag{3.8}$$

corresponding to the eigen value:

$$E = \frac{p_x^2}{2m}, \tag{3.9}$$

where E must also be a continuous variable. Thus, we reached the important conclusion that *the energy of a particle in free space is not quantized* since p_x is not quantized.

The fact that $\Psi_{p_x}(x)$ is also an eigen function of $\hat{H}$ also means that the energy and linear momentum of a particle in this state are precisely known, or both can be simultaneously measured precisely. This is expected, based on the discussions on the uncertainty principle and the commutation relationship in connection with Eqs. (2.7) and (2.8), since the Hamiltonian and the linear momentum operators commute in this case:

$$[\hat{H}, \hat{p}_x] = \left[-\frac{\hbar^2}{2m}\frac{\partial^2}{\partial x^2} \, , \, -i\hbar\frac{\partial}{\partial x} \right] = 0.$$

It is an example which shows that *when two operators commute, it is possible to find "simultaneous eigen functions" of the two*. In the case under discussion, $\Psi_{p_x}(x)$ is the simultaneous eigen function of both $\hat{H}$ and $\hat{p}_x$.

Knowing the eigen functions of the Hamiltonian also means that it is immediately possible to find the solutions of the corresponding time dependent Schrödinger equation or the time dependence of $\Psi_{p_x}(x)$ or $\Psi_E(x)$:

$$\Psi_E(x, t) = \sqrt{N}e^{\frac{i}{\hbar}p_x x - \frac{i}{\hbar}Et}, \tag{3.10}$$

according to Eq. (2.21).

Amazingly, it has the form of a plane monochromatic wave:

$$\Psi_E(x, t) = \sqrt{N}e^{ik_x x - i\omega t},$$

with the amplitude $\sqrt{N}$. The propagation constant and frequency are, respectively:

$$k_x = \frac{p_x}{\hbar}, \tag{3.11}$$

and

$$\omega = \frac{E}{\hbar}. \tag{3.12}$$

The probability distribution function $|\Psi_{p_x}(x, t)|^2 = N$ is, then, the intensity of the wave.

The state function (3.10) endows the particle with all the properties of a wave. It is a "matter wave" that is also known as the "de Broglie wave," in honor of its discoverer; hence, the state function is also a "wave function." This is the origin of the wave/particle duality, and it is an explicit consequence of the uncertainty principle (Postulate 2). The corresponding "de Broglie wavelength" is, from (3.11) and (3.9):

$$\lambda_{\mathrm{d}} = \frac{2\pi}{k_x} = \frac{h}{p_x} = \frac{h}{\sqrt{2mE}}. \tag{3.13}$$

One cannot over-emphasize the practical importance of these results, (3.8)–(3.13).

If a beam of particles traveling with a constant velocity is also a plane wave, it should be possible to observe such wave phenomena as, for example, the equivalent of the well-known Bragg diffraction or Young's two-slit interference experiment in optics. Indeed, in the well-known Davidson–Germer experiment (see, for example, Bohm (1951)), Bragg diffraction patterns of electrons scattered from various crystalline solids were observed which proved experimentally the wave nature of particles. Suitable beams of particles can, thus, be used in different ways as powerful tools to study the atomic structures of all kinds of materials and structures.

If the particle is not traveling in free space, but is in a region of constant potential energy V, then the de Broglie wavelength given in (3.13) is:

$$\lambda_{\mathrm{d}} = \frac{h}{\sqrt{2m(E - V)}}, \tag{3.14}$$

from the solution of the corresponding time-independent Schrödinger equation. Thus, for the de Broglie wave, the equivalent "relative index of refraction" of the medium is proportional to $\sqrt{(E - V)/E}$ and can be changed through the potential energy term in the Hamiltonian. For charged particles such as electrons and ions, one can then use electrostatic potentials to manipulate de Broglie waves, just like the use of optical elements such as lenses, prisms, etc. Numerically, the de Broglie wavelength of an electron with a kinetic energy of 1 eV (1 electron-volt $= 1.6 \times 10^{-12}$ erg) is 1.23 nm. By comparison, the wavelength of a photon of energy E is:

$$\lambda_{\mathrm{ph}} = \frac{2\pi}{k_x} = \frac{c}{\nu} = \frac{hc}{E}. \tag{3.15}$$

where c is the velocity of light, 3×10^{10} cm/sec. For a photon of energy 1 eV, the wavelength is 1.24 μm, which is much larger than that of the de Broglie wave of an electron of the same energy. It is not very difficult to accelerate electrons or ions to energies much greater than 1 eV and use electric and magnetic fields to manipulate and precisely control the motions of charged particles. Thus, devices based on the wave nature of particles, such as electrons or ions, potentially can have much higher resolution than optical instruments and are indeed widely used in, for example, lithography, microfabrication processes, and analytic instruments in the electronics industry and in many branches of science.

3.2 Momentum representation and wave packets

Let us now relax the condition that the linear momentum of the particle is precisely specified. Suppose the spread in the momentum value, or the uncertainty, is $\Delta p_x \neq 0$. The uncertainty principle states that the uncertainty in the position of the particle Δx can correspondingly be finite, meaning the particle can now be localized. For the position of the particle to be known precisely, the momentum must be totally uncertain. This conjugate relationship between the momentum and the position can be made more precise on the basis of Eqs. (2.28b), (2.29b), and (2.31):

$$\Psi(x) = \int \Psi(p_x)\Psi_{p_x}(x)\mathrm{d}p_x, \tag{3.16}$$

where

$$\Psi(p_x) = \int \Psi^*_{p_x}(x)\Psi(x)\mathrm{d}x \tag{3.17}$$

and

$$\int \Psi^*_{p_x}(x)\Psi_{p'_x}(x)\mathrm{d}x = \delta\left(p_x - p'_x\right). \tag{3.18}$$

To satisfy the orthonormality condition (3.18) for the continuous eigen values p_x, the eigen function $\Psi_{p_x}(x)$ must be normalized as follows:

$$\Psi_{p_x}(x) = \frac{1}{\sqrt{2\pi\hbar}}e^{\frac{i}{\hbar}p_x x}, \tag{3.19}$$

so that

$$\int_{-\infty}^{\infty} \Psi^*_{p_x}(x)\Psi_{p'_x}(x)\mathrm{d}x = \lim_{L\to\infty}\frac{1}{2\pi\hbar}\int_{-L}^{L}e^{-\frac{i}{\hbar}(p_x-p'_x)x}\mathrm{d}x$$

$$= \frac{1}{\pi}\lim_{L\to\infty}\frac{\sin\left[(p_x-p'_x)L/\hbar\right]}{(p_x-p'_x)} = \delta(p_x - p'_x).$$

(The mathematical basis for the last equality in the above equation is a little subtle. See, for example, the more detailed discussions on the Dirac delta function, following Eq. (8.20) in chapter 8, and Dirac (1947).)

Substituting (3.19) into Eq. (3.16) shows that an arbitrary state function $\Psi(x)$ can be expanded as a superposition of de Broglie waves:

$$\Psi(x) = \frac{1}{\sqrt{2\pi\hbar}} \int\limits_{-\infty}^{\infty} \Psi(p_x)\, e^{\frac{i}{\hbar}p_x x} dp_x, \tag{3.20}$$

and, from (3.17) and (3.19), the corresponding complex amplitude function $\Psi(p_x)$ is exactly the spatial Fourier transform of the state function:

$$\Psi(p_x) = \frac{1}{\sqrt{2\pi\hbar}} \int\limits_{-\infty}^{\infty} e^{-\frac{i}{\hbar}p_x x}\Psi(x) dx, \tag{3.21}$$

which is also known as the "the momentum representation" $\Psi(p_x)$ of the state function $\Psi(x)$. The corresponding x component of the de Broglie wave vector k_x is equal to $p_x/\hbar$. The probability distribution function in the momentum space is then the square of the momentum representation of this state function, $|\Psi(p_x)|^2$. It has the physical meaning that $|\Psi(p_x)|^2 dp_x$ is the probability that measurements of the momentum of the particle in this state will find the value in the range from p_x to $p_x + dp_x$.

Consider, for example, the case where the normalized probability distribution function in the momentum space is a Gaussian function with an average value $\langle p \rangle$ and uncertainty Δp:

$$|\Psi(p_x)|^2 = \frac{1}{\sqrt{2\pi(\Delta p)^2}}\, e^{-\frac{(p_x - \langle p \rangle)^2}{2(\Delta p)^2}}. \tag{3.22}$$

By choosing the reference phase factor to be zero, the momentum representation of the state function is taken to be:

$$\Psi(p_x) = \frac{1}{\sqrt[4]{2\pi(\Delta p)^2}}\, e^{-\frac{(p_x - \langle p \rangle)^2}{4(\Delta p)^2}}. \tag{3.23}$$

Taking the inverse Fourier transform according to (3.16), the corresponding state function must be:

$$\Psi(x) = \frac{1}{\sqrt[4]{2\pi(\Delta x)^2}} e^{-\frac{x^2}{4(\Delta x)^2} + \frac{i}{\hbar}\langle p \rangle x}, \tag{3.24}$$

and the normalized probability distribution function in real space is:

$$|\Psi(x)|^2 = \frac{1}{\sqrt{2\pi(\Delta x)^2}} e^{-\frac{x^2}{2(\Delta x)^2}}, \tag{3.25}$$

with the average value $\langle x \rangle = 0$ and the uncertainty spread $\Delta x^2 = [\hbar/2\Delta p]^2$.

Equations (3.22)–(3.25) lead to some very interesting conclusions. As would be expected on the basis of ordinary Fourier transform theory, if Δp^2 is finite, the probability distribution function of the position of the particle also has a finite width, meaning the particle can be localized as a "wave packet." In fact, these results show that if the probability distribution in the momentum space is Gaussian, (3.22), it is also Gaussian in the real space, (3.25), and vice versa. Also, the uncertainty product is:

$$\Delta p_x \Delta x = \frac{\hbar}{2},$$

which satisfies, and is at the minimum allowed by, Heisenberg's uncertainty relationship, (2.13). Thus, a Gaussian wave packet is a "minimum uncertainty wave packet," and is in some sense "the best one can do or the most one can know about a wave packet."

3.3 Problems

3.1 Sketch the de Broglie wavelength versus the kinetic energy up to 10 eV (= 1.6 $\times 10^{-18}$ Joules) for

(a) electrons, protons, and
(b) neutrons, and compare these results with the corresponding result for photons.

3.2 Suppose we know that there is a free particle initially located in the range $-a < x < a$ with a spatially uniform probability.

(a) Give the normalized state function $\Psi(x, t = 0)$ of the particle in the Schrödinger representation. Assume the phase of the wave function is arbitrarily chosen to be zero.
(b) Give the corresponding momentum representation of the particle.
(c) Give the corresponding state function at an arbitrary later time $\Psi(x, t > 0)$. (You can give the result in the integral form.)

3.3 Consider a free particle with the initial state function in the form of:

$$\Psi(x, t = 0) = Ae^{-ax^2 + ikx}.$$

(a) Normalize this state function.
(b) Find the corresponding momentum representation of this state function.
(c) Find the corresponding state function $\Psi(x, t > 0)$.
(d) Find the expectation values of the position and momentum, and their respective uncertainties, of the particle in this state at an arbitrary time $t > 0$.
(e) Show that Heisenberg's uncertainty principle holds for this state.

4 Particles at boundaries, potential steps, barriers, and in quantum wells

Going beyond the motion of particles in free space considered in the previous chapter, the dynamics of particles subject to various forces, or more appropriately in the language of quantum mechanics under the actions of potentials of various forms, are studied in this chapter. They include: a simple boundary, potential steps, potential barriers, and quantum wells. Even with such simple models, new and practically important quantum phenomena will show up. These include the quantum mechanical reflection and transmission effects; the quantum mechanical tunneling effect, which is the basis for the practically important tunnel diode, for example; the appearance of bound states in quantum wells, which have the same origin as the quantized energy levels in atoms, molecules, and ultimately the states of electrons, or the band structures in semiconductors; etc. The latter lead to devices such as transistors and diode lasers.

4.1 Boundary conditions and probability currents

More important than particles moving in free space are the dynamics of particles subject to various forces, or in regions of different potentials.

Let us consider first a simple one-dimensional problem. A beam of particles moving with a constant initial velocity v_0, or kinetic energy $T_0 = \frac{1}{2}mv_0^2 = \frac{p_0^2}{2m}$, in Region I is incident on a simple boundary (at $x = 0$) separating two constant potential-energy regions:

$$V(x) = \begin{cases} 0 & \text{in Region I,} \quad x < 0 \\ V_> & \text{in Region II,} \quad x > 0. \end{cases} \tag{4.1}$$

$V_>$ can be positive $(+V_0)$ or negative $(-V_0)$ corresponding to a potential step up or down. What will happen to the state of the particles passing through such a boundary between the potential regions?

First, because the probability distribution function must be single-valued everywhere in space, it must be continuous across the boundary, which implies that the wave function must be continuous:

$$\Psi^{(I)}(x = 0, t) = \Psi^{(II)}(x = 0, t). \tag{4.2}$$

If the incident wave function in Region I corresponds to a particle with a constant velocity (hence, constant energy), $\Psi^{(I)}(x, t)$ must be an eigen function $\Psi_E^{(I)}(x, t)$ of the Hamiltonian corresponding to the eigen value $E = \frac{p_0^2}{2m}$ with the time dependence $e^{-\frac{i}{\hbar}Et}$. To satisfy the condition (4.2), the wave functions on both sides of the boundary must have the same time dependence, or $\Psi^{(I)}(x, t) = \Psi_E^{(I)}(x)e^{-\frac{i}{\hbar}Et} = \Psi^{(II)}(x, t) = \Psi^{(II)}(x)e^{-\frac{i}{\hbar}Et}$ at $x=0$, with the same E, which means *the total energy must be conserved and E is a constant of motion across the boundary.*

The **continuity of wave functions across a boundary**, Eqs. (4.2), *is a most important basic boundary condition on all wave functions in quantum mechanics.*

Consider now the boundary condition on the spatial derivative of the wave function, $\frac{\partial}{\partial x}\Psi_E(x)$. Since the wave function must satisfy the time-independent Schrödinger equation, we can integrate it over an infinitesimal range δ across the boundary from $x = -\delta/2$ to $+\delta/2$:

$$\int_{-\frac{\delta}{2}}^{+\frac{\delta}{2}} \left[-\frac{\hbar^2}{2m}\frac{\partial^2}{\partial x^2} + V(x) \right] \Psi_E(x)\mathrm{d}x = E \int_{-\frac{\delta}{2}}^{+\frac{\delta}{2}} \Psi_E(x)\mathrm{d}x,$$

$$-\frac{\hbar^2}{2} \left[\frac{1}{m}\frac{\partial}{\partial x}\Psi_E(x) \right]_{x=-\frac{\delta}{2}}^{x=+\frac{\delta}{2}} = \left[E - \frac{1}{2}V_> \right] \Psi_E(x = 0)\,\delta.$$

In the limit of $\delta \to 0$, one has the boundary condition:

$$\left[\frac{1}{m_I}\frac{\partial}{\partial x}\Psi_E^{(I)}(x) \right]_{x=0} = \left[\frac{1}{m_{II}}\frac{\partial}{\partial x}\Psi_E^{(II)}(x) \right]_{x=0}. \tag{4.3}$$

Since an arbitrary wave function can always be expanded as a superposition of the eigen functions of the Hamiltonian, $\Psi(x, t) = \sum_n C_n \Psi_{E_n}(x)e^{-\frac{i}{\hbar}E_n t}$, the corresponding derivatives of any linear combination of the eigen functions must satisfy the same condition, or:

$$\left[\frac{1}{m_I}\frac{\partial}{\partial x}\Psi^{(I)}(x, t) \right]_{x=0} = \left[\frac{1}{m_{II}}\frac{\partial}{\partial x}\Psi^{(II)}(x, t) \right]_{x=0}. \tag{4.4}$$

The $(1/m_{I,II})$ factors in (4.4) need particular attention. ***In the case where the "effective" mass of the particle does not change across the boundary, the derivative of the wave function must be continuous across the boundary:***

$$\left[\frac{\partial}{\partial x}\Psi^{(I)}(x, t) \right]_{x=0} = \left[\frac{\partial}{\partial x}\Psi^{(II)}(x, t) \right]_{x=0}. \tag{4.5}$$

The concept of "effective mass" of particles in solids, semiconductors in particular, will be discussed in detail in Chapter 10. For the purpose of the present discussion, it can be considered as the "mass" of the particles under consideration. It is important to note that, *when the effective mass of the particle changes across the boundary between* *two potential regions, such as across a heterojunction between two different semiconductors (e.g. GaAs and Al$_x$Ga$_{1-x}$As), it is **the spatial derivative of the wave function divided by the effective mass that must be continuous across the boundary, as in** (4.4).*

We will now show that the boundary condition on the derivative of the wave function is also a consequence of the physical condition that the "probability current," $\vec{J}$, or the number of particles through a boundary per unit area per unit time, must be continuous, because particles cannot accumulate, or be stored, in an infinitely thin boundary. The probability current can be defined in terms of the wave function on the basis of the usual mass-conservation equation:

$$\nabla \cdot \vec{J} = -\frac{\partial}{\partial t}\rho = -\frac{\partial}{\partial t}|\Psi(\vec{r}, t)|^2. \tag{4.6}$$

In the one-dimensional case,

$$\frac{\partial J_x}{\partial x} = -\Psi^*(x, t)\frac{\partial}{\partial t}\Psi(x, t) - \Psi(x, t)\frac{\partial}{\partial t}\Psi^*(x, t). \tag{4.7}$$

Making use of the time-dependent Schrödinger equation (2.16), the right side of (4.7) leads to:

$$\frac{\partial J_x}{\partial x} = -\frac{i\hbar}{2m}\Psi^*(x, t)\frac{\partial^2}{\partial x^2}\Psi(x, t) + \frac{i\hbar}{2m}\Psi(x, t)\frac{\partial^2}{\partial x^2}\Psi^*(x, t)$$
$$= \frac{\partial}{\partial x}\left[-\frac{i\hbar}{2m}\Psi^*(x, t)\frac{\partial}{\partial x}\Psi(x, t) + \frac{i\hbar}{2m}\Psi(x, t)\frac{\partial}{\partial x}\Psi^*(x, t)\right],$$

and the x component of the probability current can, thus, be defined as:

$$J_x = -\frac{i\hbar}{2m}\Psi^*(x, t)\frac{\partial}{\partial x}\Psi(x, t) + \frac{i\hbar}{2m}\Psi(x, t)\frac{\partial}{\partial x}\Psi^*(x, t). \tag{4.8}$$

Finally, generalizing to three dimensions, we have an expression for the general **probability current**:

$$\vec{J} = -\frac{i\hbar}{2m}\Psi^*(\vec{r}, t)\nabla\Psi(\vec{r}, t) + \frac{i\hbar}{2m}\Psi(\vec{r}, t)\nabla\Psi^*(\vec{r}, t). \tag{4.9}$$

Assuming continuity of the current as defined in (4.9) and the continuity of the wave function across the boundary, (4.2), one obtains the boundary condition on the derivative of the wave function (4.5) again. Note also the ($1/m$) factor in the definition of the current in (4.9). If the mass of the particle changes across a boundary, the continuity condition on the current leads to the boundary condition (4.4), instead of (4.5).

Note that, if the wave function Ψ is such that $\Psi^*(\vec{r}, t)\nabla\Psi(\vec{r}, t)$ is purely real (no imaginary part), (4.9) shows that the current $\vec{J}$ must be equal to zero. This means that, for the current in a given region not to be zero, the corresponding de Broglie wave in that region must be some sort of a propagating wave. For exponentially damped waves or standing waves, the corresponding particle current is zero, as is intuitively expected.

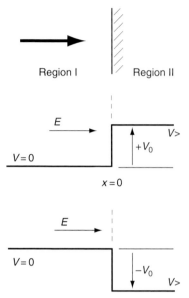

Figure 4.1. A beam of particles incident on a potential-energy step.

4.2 Particles at a potential step, up or down

We now ask what happens to a beam of particles moving with a constant velocity impinging on the potential step at $x = 0$ as specified in Eq. (4.1), where the potential-energy step can be up or down depending on whether $\Delta V \equiv V_>$ is positive ($+V_0$) or negative ($-V_0$) (Figure 4.1)? Such a boundary can represent the interface, or a heterojunction, between two material regions and plays an extremely important role in modern solid state electronic and laser devices. For the present discussion, we assume that any change in the mass of the particles across the boundary is negligible. In semiconductor optical structures, the consequence of the change in the effective mass of charge-carriers across some practical heterojunctions often may not be negligible.

The answer according to *classical mechanics* based only on energy considerations is: if $\Delta V = -V_0$ is negative, the particles will have 100% probability of going from Region I to Region II while gaining in kinetic energy of the amount V_0 and in velocity from v_0 to $\sqrt{v_0^2 + 2V_0/m}$. If $\Delta V = +V_0$ and $T_0 = 1/2\ mv_0^2 < V_0$, then there is 100% probability that no particle will get into Region II. If $\Delta V = +V_0$ and $T_0 = 1/2\ mv_0^2 > V_0$, the particles will again have 100% probability of going through the boundary from Region I to Region II but losing in kinetic energy from T_0 to $T_0 - V_0$ and in velocity from v_0 to $\sqrt{v_0^2 - 2V_0/m}$.

Quantum mechanics, on the other hand, paints a more detailed and intricate picture. To know exactly what will happen to the particles, it is necessary to find the wave function of the particles from Region I to II. Knowing the conditions the wave function must satisfy at the boundary, it is now possible to find the wave function throughout Regions I and II with the specified potential, (4.1). Because energy is conserved, the total energy is a constant of motion and the same in both regions; therefore, the wave function must be an eigen function of the Hamiltonian of the particle corresponding to the eigen value $E = \frac{1}{2}mv_0^2$:

$$\hat{H}\Psi_E(x) = E\Psi_E(x).$$

In the Schrödinger representation, it is:

$$\left[-\frac{\hbar^2}{2m}\frac{\partial^2}{\partial x^2} + V(x)\right]\Psi_E(x) = E\Psi_E(x), \tag{4.10}$$

where $V(x)$ is given in (4.1).

The general solution corresponding to the situation where a beam of $|A|^2$ particles per unit length is incident on the boundary at $x = 0$ from Region I is:

$$\Psi_E^{(I)}(x) = Ae^{ik_1x} + Be^{-ik_1x}, \qquad \text{for } x < 0, \tag{4.11a}$$

and

$$\Psi_E^{(II)}(x) = Ce^{ik_2x}, \qquad \text{for } x > 0, \tag{4.11b}$$

where

$$k_1 = \frac{\sqrt{2mE}}{\hbar}, \tag{4.12a}$$

$$k_2 = \frac{\sqrt{2m(E - V_>)}}{\hbar}. \tag{4.12b}$$

In general, when E is less than V, it is also useful to define explicitly the imaginary part of k as α:

$$k = i\alpha \qquad \text{and} \qquad \alpha = \frac{\sqrt{2m(V - E)}}{\hbar}. \tag{4.12c}$$

Note that the solutions are of the same forms as the solutions of the wave equations for electromagnetic or optical waves propagating in two regions of different dielectric constants. Here the kinetic energy $(E - V)$ for the de Broglie waves plays the role of the relative dielectric constant, in complete analogy with electromagnetic or optical waves. A, B, and C in Eqs. (4.11a & b) are the amplitude of the incident, reflected, and transmitted de Broglie waves, respectively. Applying the boundary conditions (4.2) and (4.5) to (4.11a) and (4.11b), we have:

$$A + B = C$$

$$k_1(A - B) = k_2C;$$

therefore,

$$\frac{B}{A} = \frac{k_1 - k_2}{k_1 + k_2} = \frac{\sqrt{E} - \sqrt{E - V_>}}{\sqrt{E} + \sqrt{E - V_>}}, \tag{4.13a}$$

$$\frac{C}{A} = \frac{2k_1}{k_1 + k_2} = \frac{2\sqrt{E}}{\sqrt{E} + \sqrt{E - V_>}}. \tag{4.13b}$$

B/A and C/A are the ratios of the complex amplitude of the reflected and transmitted waves to that of the incident wave, respectively. With these, we can now find the transmission coefficient T and the reflection coefficient R defined as the ratios of the corresponding currents, from Eq. (4.8):

$$T \equiv \frac{J_x^{(\text{transmitted})}}{J_x^{(\text{incident})}} \equiv \frac{k_2}{k_1} \left| \frac{C}{A} \right|^2 = \left| \frac{4k_1 k_2}{(k_1 + k_2)^2} \right| = \left| \frac{4\sqrt{E(E - V_>)}}{(\sqrt{E} + \sqrt{E - V_>})^2} \right|, \tag{4.14a}$$

for E greater than $V_>$ or k_2 purely real. For E less than $V_>$, or k_2 purely imaginary, Eq. (4.8) shows that T *must always be zero.* (E is always positive or k_1 is always purely real for a propagating wave in Region I). Also, whether E is greater or less than $V_>$, or whether k_2 is purely real or imaginary, the reflection coefficient is always:

$$R \equiv \frac{J_x^{(\text{reflected})}}{J_x^{(\text{incident})}} \equiv \left| \frac{B}{A} \right|^2 = \left| \frac{k_1 - k_2}{k_1 + k_2} \right|^2 = \left| \frac{\sqrt{E} - \sqrt{E - V_>}}{\sqrt{E} + \sqrt{E - V_>}} \right|^2. \tag{4.14b}$$

These results have interesting and important physical implications.

First, as can be shown from (4.14a) and (4.14b), the total current into and out of the boundary is always conserved as expected, or $R + T = 1$, regardless of whether E is greater or less than $V_>$, as expected since no particle can accumulate in the thin boundary layer.

Second, (4.14b) shows that, for a finite potential step, whether it is down or up ($\Delta V = -$ or $+ V_0$), R can never be equal to zero, or the particles can never have 100% probability of going through the sharp boundary ($T = 1$) separating the two regions, unlike in the classical case. According to classical mechanics, if the initial kinetic energy of the incident particles is greater than ΔV, there will be no reflection. According to quantum mechanics, this reflection is a wave phenomenon. As such, whether such a reflection takes place or not will depend on how 'sharp' the 'boundary' is. Reflection at the boundary can only occur when the boundary is sharp relative to the wavelength $\lambda_d = \frac{h}{\sqrt{2mE}}$ of the incident de Broglie wave. Thus, no such reflection can possibly be seen experimentally in the macroscopic world, because no physical potential can vary spatially fast enough to appear as a sharp boundary to a moving macroscopic particle of any measurable mass and velocity.

Third, when E is less than $V_>$, the corresponding wave function is a damped wave in Region II:

$$\Psi_E^{(\text{II})}(x) = Ce^{-\alpha_2 x}, \qquad \text{for } x > 0, \tag{4.15a}$$

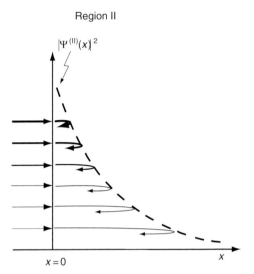

Region II

Figure 4.2. Schematic showing that the probability of penetration of the particles decreases with the distance into Region II.

where

$$\alpha_2 = \frac{\sqrt{2m(V_> - E)}}{\hbar}. \tag{4.15b}$$

Based on the discussion following (4.9), there can be no net current flow in Region II. Thus, $T = 0$ and $R = 1$. Nevertheless, the probability distribution function of the particles in Region II is not equal to zero and, hence, there are particles present in Region II. It can only mean that, in Region II, there are as many particles going in the $+x$ as the $-x$ direction, or every particle that penetrates into Region II eventually turns around and heads back into Region I (Figure 4.2); thus $R = 1$ and $T = 0$. The fact that the probability distribution function is of the form $\left|\Psi_E^{(II)}\right|^2 = |C|^2 e^{-2\alpha_2 x}$ means that the number of particles in Region II decreases exponentially with distance from the boundary $x = 0$, which is a quantum mechanic effect.

How does one know such an effect actually takes place on the atomic and subatomic scale? This effect can manifest itself in a number of ways where it can be studied in the macroscopic world, as will be discussed in more detail in later chapters. One possible simple situation where this effect can be seen directly is the tunneling effect. Suppose, for example, there is a second interface separating Region II and another potential Region III, and $V_{II} > E > V_I$ and V_{III} (see, for example, Figure 4.3). Because the wave function must be continuous at all boundaries, some particles reaching the second boundary between Regions II and III will have a finite probability of passing through the boundary and reaching Region III and being detected. This is called the "quantum mechanical tunneling effect." With the current semiconductor microfabrication technology, it is relatively easy to fabricate a structure that can be used to observe and study such a tunneling effect in detail. The basic principle of the

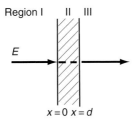

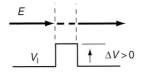

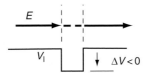

Figure 4.3. A beam of particles traveling at a constant velocity incident on a potential barrier in the region $0 < x < d$.

phenomenon of tunneling through a potential barrier will be developed in the following section.

4.3 Particles at a barrier and the quantum mechanical tunneling effect

From a single boundary separating two potential regions of semi-infinite extent, we now move on to a finite potential structure:

$$V(x) = \begin{cases} V_{\mathrm{I}} = V_{\mathrm{III}} & \text{for} \quad 0 < x \quad \text{and} \quad x > d \\ V_{\mathrm{II}} = V_{\mathrm{I}} + \Delta V & \text{for} \quad 0 < x < d, \end{cases} \tag{4.16}$$

consisting of two boundaries separating three potential regions, as shown in Figure 4.3. A new key feature that will show up in this structure is the quantum mechanical tunneling effect.

For the three-region case, the general form of the solution of the corresponding time-independent Schrödinger equation is:

$$\Psi_E^{(\mathrm{I})}(x) = A\mathrm{e}^{ik_1 x} + B\mathrm{e}^{-ik_1 x}, \qquad \text{for } x < 0, \tag{4.17a}$$

$$\Psi_E^{(\mathrm{II})}(x) = C\mathrm{e}^{ik_2 x} + D\mathrm{e}^{-ik_2 x}, \qquad \text{for } 0 < x < d, \tag{4.17b}$$

$$\Psi_E^{(III)}(x) = Fe^{ik_3 x}, \qquad \text{for } x > d, \tag{4.17c}$$

where

$$k_1 = \frac{\sqrt{2m(E - V_{\mathrm{I}})}}{\hbar}, \tag{4.18a}$$

$$k_2 = \frac{\sqrt{2m(E - V_{\mathrm{II}})}}{\hbar}, \tag{4.18b}$$

$$k_3 = \frac{\sqrt{2m(E - V_{\mathrm{III}})}}{\hbar}. \tag{4.18c}$$

In this section, we will consider only the cases where $E > V_{\mathrm{I}} = V_{\mathrm{III}}$. In that case, k_1 and k_3 are always real. k_2, on the other hand, can be real or imaginary. In the latter case, we define the imaginary part of k_2 as:

$$k_2 = i\alpha_2 \quad \text{and} \quad \alpha_2 = \frac{\sqrt{2m(V_{\mathrm{II}} - E)}}{\hbar}, \tag{4.18d}$$

and

$$\Psi_E^{(II)}(x) = Ce^{-\alpha_2 x} + De^{\alpha_2 x}, \qquad \text{for } 0 < x < d. \tag{4.17d}$$

Applying the boundary conditions (4.2) and (4.5) to (4.17a–c) at $x = 0$ and d gives, after some algebra:

$$\frac{F}{A} = \frac{e^{-ik_3 d}}{\left[\cos k_2 d - i \frac{k_1^2 + k_2^2}{2k_1 k_2} \sin k_2 d\right]}, \tag{4.19}$$

and the corresponding current transmission coefficient is:

$$T = \left[1 + \frac{\Delta V^2 \sin^2 k_2 d}{4(E - V_{\mathrm{I}})(E - V_{\mathrm{II}})}\right]^{-1}, \tag{4.20a}$$

for E either greater or less than V_{II}. When E is less than V_{II}, it is sometimes useful to replace the sine-function in (4.20a) by the corresponding sinh-function:

$$T = \left[1 + \frac{\Delta V^2 \sinh^2 \alpha_2 d}{4(E - V_{\mathrm{I}})(V_{\mathrm{II}} - E)}\right]^{-1}, \tag{4.20b}$$

so that all the factors in (4.20b) are positive real. The two forms are, however, completely equivalent and either form can be used for E either greater than V_{II} or less than V_{II}. Numerical examples of the transmission coefficient as a function of the incident energy of the particle are shown in Figure 4.4. Such traces have a number of general features that illustrate a number of important points.

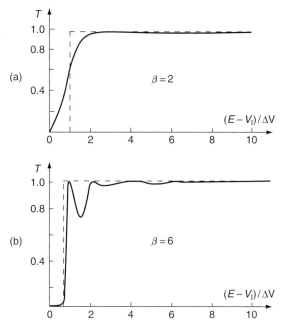

Figure 4.4. Examples of transmission curves (solid curves) of a potential barrier. (a) $\beta = \sqrt{2m\Delta V}d/\hbar = 2$; (b) $\beta = 6$. The fact that T is not equal to zero for $E < V_{\mathrm{II}}$ corresponds to the quantum mechanical tunneling effect. The dashed lines show the classical limits.

First, as is obvious from Fig. 4.4(b), when the kinetic energy $E - V_{\mathrm{I}}$ in Region I is greater than the potential step ΔV, T is a damped oscillatory function of E and asymptotically approaches the value 1. The interesting point is that there are particular values of E corresponding to where $k_2 d$ is an integral multiple of π at which the transmission coefficient is also equal to unity. At these values, even though there is a barrier present, the particles have 100% probability of going through the barrier as if it were transparent. This resonance effect in the transmission is due to the constructive interference of the de Broglie waves in the forward direction as a result of multiple reflections between the two boundaries of Region II. At other values of energy, T is not equal to 1; therefore, there is always reflection back into Region I even when $E - V_{\mathrm{I}}$ is greater than the potential step ΔV. According to classical mechanics based on energy considerations only, when $E - V_{\mathrm{I}}$ is greater than ΔV, every particle will always go over the barrier into Region III.

Second, when $E - V_{\mathrm{I}}$ is less than the potential step ΔV, according to classical mechanics, no particle should get into Region II, much less into Region III. Yet (4.20a or b) and Figure 4.4 show that the transmission from Region I through Region II into Region III is finite. It means that, even though the kinetic energy of the particles in Region II is negative, the particles nevertheless have a finite probability of "tunneling" through the barrier Region II and emerging in Region III. This quantum mechanical tunneling effect is the basis for many physical phenomena and has many important practical applications in electronics, such as the tunnel diode,

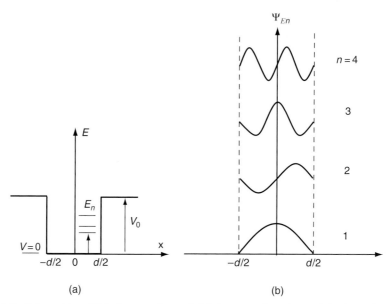

(a)   (b)

Figure 4.5. Schematics of (a) the energies and (b) the wave functions of the bound states in a square well potential or quantum well. (The horizontal axes for the wave functions in (b) are shifted for clarity.)

cold emission of electrons from metals, Josephson superconductor tunneling, etc. Mathematically, it should be noted that, when $E - V_{\mathrm{I}}$ is less than the potential step ΔV, even though the wave function in the barrier region consists of superpositions of non-propagating waves of the forms $\mathrm{e}^{\pm \alpha x}$, there still is a net probability current flowing through the barrier. This is because the amplitudes of these non-propagating waves are complex because of the boundary conditions and, therefore, the total wave function in the barrier region is not purely real. As a result, the total probability current, from (4.9), in the barrier region is not equal to zero.

4.4 Quantum wells and bound states

The potential structure of special interest, as shown in Figure 4.5, is the case where the potential-energy step $\Delta V = -V_0$ and $E - V_{\mathrm{I}} = E - V_{\mathrm{III}}$ are both negative but $E - V_{\mathrm{II}}$ is positive. For this case, we shift the origin of the energy scale to the bottom of the potential-well, as shown in the figure. The potential-energy function is, therefore:

$$V(x) = \begin{cases} V_0 & \text{for} & |x| > d/2, \\ 0 & \text{for} & |x| < -d/2. \end{cases} \qquad (4.21)$$

This is the case where the kinetic energy of the particle in both Regions I and III is negative and the corresponding de Broglie waves must be damped non-propagating waves. Therefore, the particle is trapped in, or "bound" to, Region II with a finite probability of penetrating a small distance into and then turning around in the wall regions defined by Regions I and III. As will be shown below, the energy of the particles in the well must be quantized. These are the single-particle quantized "bound states" of the square well potential, or the "quantum well."

Infinite potential well

We begin with the limiting case of a particle confined in an infinite potential well, $V_0 \to \infty$; or Regions I and III represent impenetrable walls. In this case, the wave functions in Regions I and III must vanish. This follows from the fact that, wherever V is infinite, the wave function must be zero in order to satisfy Schrödinger's equation. Thus, the solution of the time-independent Schrödinger equation in this case is of the form:

$$\Psi_{E_n}^{(I,III)} = 0, \qquad\qquad \text{for} \qquad |x| > d/2,$$
$$\Psi_{E_n}^{(II)} = Ae^{ik_2x} + Be^{-ik_2x}, \qquad \text{for} \qquad |x| < d/2, \tag{4.22}$$

where, with reference to Figure 4.5:

$$k_2 = \frac{\sqrt{2mE}}{\hbar}. \tag{4.23}$$

To satisfy the boundary condition (4.2) at $x = d/2$ and $-d/2$:

$$Ae^{ik_2\frac{d}{2}} + Be^{-ik_2\frac{d}{2}} = 0,$$
$$Ae^{-ik_2\frac{d}{2}} + Be^{ik_2\frac{d}{2}} = 0, \tag{4.24a}$$

which implies that

$$e^{2ik_2d} = 1, \qquad \text{or} \quad e^{ik_2d} = \pm 1. \tag{4.24b}$$

Thus, k_2 must have discrete values and is equal to integral multiples of π:

$$k_{2n}d = n\pi, \quad \text{where} \quad n = 1, 2, 3, 4, \ldots, \tag{4.25}$$

and, based on the interpretation according to (3.11), the corresponding momentum of the particle inside the well must have quantized values of:

$$p_{xn} = \pm\frac{n\pi\hbar}{d}, \qquad\qquad \text{where} \qquad n = 1, 2, 3, 4, \ldots \tag{4.25a}$$

One should be careful with this interpretation, however, because of the finite range of x over which the wave function is defined due to the finite width of the well. If the momentum of the particle in the box is to be measured, the uncertainty principle will come into play. Based on the representation given in (3.21), the probability distribution function of the measured values of the momentum in the box in terms of the free-space eigen function of the momentum operator of the particle moving in each direction will have an uncertainty range of the order of $\sim \hbar/d$ centered around each of the quantized values given in (4.25a), because of the finite range of x over which the wave function is defined. It will only go to zero for large d. The overall uncertainty of the measured momentum of the particle in the box based on the quantized momentum states is, on the other hand, due to the opposite directions, or the plus and minus signs, of the momentum values given in (4.25a) and has the magnitude $\left[\dfrac{n\pi\hbar}{d}\right]$, which increases with n and decreases with increasing d. (For a more detailed discussion, see, for example, Cohen-Tannoudji, *et al.* (1977) Vol. I, p. 270–274.)

The condition (4.25) leads to the important result that *the energy of the particle in the well must be quantized.* The energy eigen values are, from (4.23) and (4.25):

$$E_n = \frac{\pi^2 n^2 \hbar^2}{2md^2}, \tag{4.26}$$

unlike the energy E of the particles in free space considered in Sections 4.1 to 4.3, which is a continuous variable. Furthermore, from (4.24a):

$$
\begin{aligned}
A &= B, && \text{for} && k_{2n}d = n\pi, && \text{where } n = 1,\ 3,\ 5,\dots \\
A &= -B, && \text{for} && k_{2n}d = n\pi, && \text{where } n = 2,\ 4,\ 6,\dots
\end{aligned}
\tag{4.27}
$$

Since the particle and the corresponding wave function are now confined in a finite range $-d/2 < x < d/2$, the normalization condition is such that the integral of the corresponding probability distribution function over this range is equal to 1. Thus, the normalized energy eigen states are:

$$\Psi^{(\mathrm{II})}_{E_n}(x) = \begin{cases} \sqrt{\dfrac{2}{d}}\,\cos\dfrac{n\pi}{d}x, & \text{for} \quad n = 1, 3, 5, \dots \\[2mm] \sqrt{\dfrac{2}{d}}\,\sin\dfrac{n\pi}{d}x, & \text{for} \quad n = 2, 4, 6, \dots, \end{cases} \tag{4.28}$$

as shown schematically in Figure 4.5(b).

Note that there is no state with the quantum number $n = 0$, because the wave function $\Psi_{E_0}(x) = 0$. This implies that the lowest energy state is not when the particle is completely at rest in the box, unlike in the macroscopic world, where one can surely have a particle sitting motionless in a box. On the atomic and subatomic scale, however, if we know that the particle is confined in the range $-d/2 < x < d/2$, or the uncertainty in its position is finite, Heisenberg's uncertainty principle, (2.13), predicts that the uncertainty in the momentum of the particle must also be finite and the particle must have a minimum amount of kinetic energy. Therefore, the lowest energy the particle in the box can have must not be zero.

This lowest energy state is a stationary state corresponding to the situation where the corresponding de Broglie waves traveling in the opposite directions have a definite phase relationship such that the zeros of the corresponding interference pattern occur exactly at the boundaries $x = d/2$ and $-d/2$. At the same time, it also describes a state in which the particle continuously bounces back with the average velocities $v_x = \pm\frac{\pi\hbar}{md}$ between the two confining walls at $x = d/2$ and $-d/2$ with no energy loss. Numerically, such a quantum effect can not possibly be seen in the macroscopic world for any measurable values of mass m and square potential well width d. For example, even for

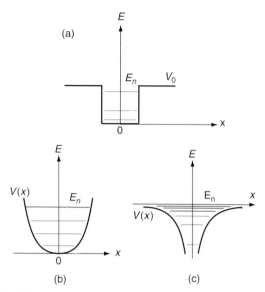

Figure 4.6. Schematic diagrams showing the quantized energy levels of (a) a square well potential, (b) potential for a harmonic oscillator, and (c) Coulomb potential for a one-electron atom.

a particle as small as one microgram and potential well width on the order of the width of a hair (tenth of a millimeter), E_1 is still only of the order of 6×10^{-44} erg and the minimum velocity is of the order of 3×10^{-19} cm/s, which are too small to be measured.

It is also of interest to note that the forms of the eigen functions, (4.28), show that the more nodes or wiggles there are in a wave function, the higher is the corresponding quantized energy of the bound state. This general feature is common to all the quantized bound states of atomic systems.

Equation (4.26) shows that the quantized energy levels increase quadratically with the quantum number n (Figure 4.6a) and decrease with increasing square of the width of the square well d^2. These general features provide a qualitative clue as to what might be expected of the pattern of quantized energy levels in potential wells of other shapes. For example, if the width of the potential well is not a constant but increases with energy, one would then expect the energy levels not to increase as fast as n^2. Indeed, if the well width increases quadratically with energy, as in the case of the harmonic oscillator, the n-squared dependence exactly cancels out the $1/d$-squared dependence. The resulting quantized energy levels of the harmonic oscillator are exactly equally spaced (see Figure 4.6b), as will be shown in more detail in Chapter 5. Also, if the well width increases even faster with energy, as in the case of the Coulomb potential in a one-electron atom where $V(r) = -e^2/r$, the quantized energy levels will actually become closer and closer together and increase as $(-1/n^2)$ (see Figure 4.6c), as will be shown in Chapter 6.

Finally, the form of the wave functions in (4.28) illustrates an important general property of the eigen functions of the Hamiltonian called *"parity"*.

The concept of parity

Note that the eigen functions $\Psi_{E_n}(x)$ in (4.28) are either symmetric or antisymmetric in x with respect to the center of the well, depending on the n value:

$$\Psi_{E_n}^{(II)}(-x) = \begin{cases} \Psi_{E_n}^{(II)}(x) & \text{for} & n = 1, 3, 5, \ldots \\ -\Psi_{E_n}^{(II)}(x) & \text{for} & n = 2, 4, 6, \ldots \end{cases} \tag{4.29}$$

The property that specifies whether the wave function changes sign or not when the coordinate axes are inverted is called the "parity" of the wave function. If the wave function does not change or only changes its sign under coordinate inversion, it is said that "the parity of the wave function is well defined." If the wave function changes more than just its sign under coordinate inversion, then the parity of the wave function is not well defined. If the parity is well defined and the wave function does not change sign and remains invariant when the coordinate axes are inverted, it is said to have even parity, such as the case with $n = 1, 3, 5, \ldots$ in (4.29). If it changes sign but otherwise remains the same, it has odd parity, such as the case with $n = 2, 4, 6, \ldots$ in (4.29).

The fact that the parity of the wave function for the square well potential is well defined is no accident. It has to do with the fact that the potential energy function in the Hamiltonian is symmetric under coordinate inversion, or more precisely $V(x)$ given in (4.21) is symmetric with respect to x, or $V(x)$ is equal to $V(-x)$. When the potential well is physically symmetric with respect to inversion of any coordinate axis, it is obvious that the probability distribution of the particle position must also be symmetric. Thus, the probability distribution function in any stationary state or the square of the eigen function of the Schrödinger equation must be symmetric:

$$|\Psi_{E_n}(-x)|^2 = |\Psi_{E_n}(x)|^2,$$

for all values of x. It follows that the wave function itself must be either symmetric or antisymmetric:

$$\Psi_{E_n}(-x) = \pm\Psi_{E_n}(x). \tag{4.30}$$

We can also formally define a "parity operator $\hat{\mathcal{P}}$," which means the process of determining the parity of the wave function. Consistent with the meaning of operators, as discussed in Section 3.2:

$$\hat{\mathcal{P}}\Psi_{E_n}(x) = \begin{cases} \Psi_{E_n}^{(II)}(x) & \text{for} & n = 1, 3, 5, \ldots \\ -\Psi_{E_n}^{(II)}(x) & \text{for} & n = 2, 4, 6, \ldots \end{cases} \tag{4.31}$$

This means that *the parity operator must have the eigen value* $+1$ *with the eigen functions* $\Psi_{E_n}(x)$ where $n = 1, 3, 5, \ldots$, and the eigen value -1 with the eigen functions

$\Psi_{E_n}(x)$ where $n=2, 4, 6, \ldots$, respectively. From (4.30) and (4.31), the parity operator can also be interpreted as the operation of inverting the coordinate axis:

$$\hat{\mathcal{P}}\Psi_{E_n}(x) = \Psi_{E_n}(-x).$$ (4.32)

Furthermore, since $\Psi_{E_n}(x)$ in this case is a simultaneous eigen function of both the parity operator and the Hamiltonian, the product $(\hat{H}\hat{\mathcal{P}})$ operating on any wave function must be equal to the effect of $(\hat{\mathcal{P}}\hat{H})$ on the same wave function:

$$\hat{H}\hat{\mathcal{P}}\sum_n C_n\Psi_{E_n}(x) = \pm\sum_n C_n E_n\Psi_{E_n}(x) = \hat{\mathcal{P}}\hat{H}\sum_n C_n\Psi_{E_n}(x).$$

This means the parity operator must commute with the Hamiltonian if the potential energy is symmetric under inversion of the coordinate axis:

$$[\hat{H}, \hat{\mathcal{P}}] = 0,$$ (4.33)

since the kinetic energy term in the Hamiltonian is always symmetric under inversion of the coordinate axes. Generalizing the above discussion to the three-dimensional case, we have the important conclusion that *if the Hamiltonian is invariant under inversion of the coordinate axes, the parity of the system is well defined and the Hamiltonian commutes with the parity operator, meaning that the energy and parity of the system can be known and specified precisely simultaneously.* Parity plays an important role in optical transitions in atomic systems, as will be discussed in later chapters.

Extending to two and three-dimensional systems, the Hamiltonian may be invariant under other symmetry operations, such as reflection in a plane, rotation of fixed angles about an axis, or combination of these operations among themselves or any of these with the inversion operation. Considerations of the related transformation properties of the eigen functions of the Hamiltonian under such symmetry operations can reveal important information about the structural and dynamic properties of atomic and subatomic systems. It is a powerful technique that forms the basis of space group theory in atomic, molecular, and solid state physics.

Finite potential well

Suppose the depth of the potential well is not infinite, or V_0 as shown in Figure 4.5 is finite. The wave functions in the three regions are then:

$$\begin{aligned}
\Psi_{E_n}^{(I)} &= Ce^{\alpha x}, & \text{for} && x < -d/2, \\
\Psi_{E_n}^{(II)} &= Ae^{ik_2 x} + Be^{-ik_2 x}, & \text{for} && |x| < d/2, \\
\Psi_{E_n}^{(III)} &= De^{-\alpha x}, & \text{for} && x > +d/2,
\end{aligned}$$ (4.34)

where

$$\alpha = \frac{\sqrt{2m(V_0 - E)}}{\hbar} \quad \text{and} \quad k_2 = \frac{\sqrt{2mE}}{\hbar}.$$ (4.35)

To determine the quantized energy levels for the bound states with $E_n < 0$, we must apply the boundary conditions (4.2) and (4.5) to the wave functions at $x = \pm d/2$. It will lead to four coupled homogeneous algebraic equations for the four unknowns A, B, C, and D. The corresponding secular determinant of the four homogeneous algebraic equations will determine the quantized energies of the particle in the quantum well. The algebra can, however, be significantly simplified by considering the parity of the eigen functions first. Since the finite potential well under consideration is symmetric with respect to inversion of the x-axis, the parity of the eigen functions must be well defined. The eigen functions must be either symmetric or anti-symmetric under inversion of the x-axis. Let us consider the two types of eigen functions separately.

First, for the eigen functions with even parity, or the symmetric states, $C = D$ and $A = B$ in (4.34), the wave functions in the three regions of space are:

$$
\begin{aligned}
\Psi_{E_n}^{(I)} &= Ce^{\alpha x}, && \text{for} && x < -d/2, \\
\Psi_{E_n}^{(II)} &= 2A\cos k_2 x, && \text{for} && |x| < d/2, \\
\Psi_{E_n}^{(III)} &= Ce^{-\alpha x}, && \text{for} && x > +d/2.
\end{aligned}
\tag{4.36}
$$

The wave function in Region II is a coherent superposition state of two counter-propagating de Broglie waves of equal amplitude and the same phase at the middle of the well at $x = 0$. Applying the boundary conditions (4.2) and (4.5) at $x = d/2$ gives:

$$
\begin{aligned}
2A\cos(k_2 d/2) &= Ce^{-\alpha d/2}, \\
2Ak_2 \sin(k_2 d/2) &= C\alpha e^{-\alpha d/2}.
\end{aligned}
$$

The secular equation of these coupled homogeneous equations for the two unknowns A and C is:

$$
(k_2 d/2)\tan(k_2 d/2) = \alpha d/2.
\tag{4.37}
$$

Note that (4.35) also shows that:

$$
(\alpha d)^2 + (k_2 d)^2 = \frac{2mV_0 d^2}{\hbar^2}.
\tag{4.38}
$$

Simultaneous solutions of (4.37) and (4.38) will give the allowed quantized values of the momentum, $\hbar k_{2n}$, and, hence, the energies E_n from (4.35) of the symmetric (with even parity) bound states ($E_n < V_0$) of the particle in the finite potential well.

For the anti-symmetric states, or the eigen states with odd parity, $C = -D$ and $A = -B$ in (4.33). The corresponding wave functions in the three regions of space are:

$$
\begin{aligned}
\Psi_{E_n}^{(I)} &= Ce^{\alpha x}, && \text{for} && x < -d/2, \\
\Psi_{E_n}^{(II)} &= 2iA\sin k_2 x, && \text{for} && |x| < d/2, \\
\Psi_{E_n}^{(III)} &= -Ce^{-\alpha x}, && \text{for} && x > +d/2.
\end{aligned}
\tag{4.39}
$$

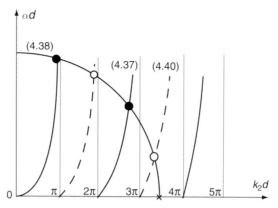

Figure 4.7. Schematic of semi-graphical solutions of Eqs. (4.37), (4.38), and (4.40). The crossing points of the quarter-circle (4.38), the tangent-like curves corresponding to the left side of Eq. (4.37), and the minus-cotangent-like curves corresponding to the left side of Eq. (4.40) give the symmetric (solid circle) and anti-symmetric (open circle) bound states, respectively.

The wave function in Region II in this case is a superposition of two counter-propagating de Broglie waves of equal amplitude but opposite phase that produces a null at the middle of the well, $x = 0$. Again, applying the boundary conditions at $x = d/2$ leads to the secular equation for the two coupled homogeneous equations for the two unknowns C and A:

$$- (k_2 d/2) \cot(k_2 d/2) = \alpha d/2. \tag{4.40}$$

Simultaneous solutions of (4.38) and (4.40) give the allowed values of the momentum, $\hbar k_{2n}$, and the quantized energies, k_n, of the anti-symmetric bound states of the particle in the finite potential well.

For quantitative results, one should obviously solve these equations numerically. To gain insights into the general characteristics of the quantized energies and the corresponding eigen functions of the bound states of the particle in the quantum well, it is useful to use a semi-graphical approach to analyze the problem. Equations (4.37), (4.38), and (4.40) are shown schematically as functions of αd versus $k_2 d$ in Figure 4.7. Qualitatively, all the features of quantized energy levels and the wave functions can be understood on the basis of such an analysis. Note that the trajectory of (4.38) is a circle. Where it crosses the trajectories of (4.37) and (4.40) gives the quantized values of $k_{2n} d$ and, thus, from (4.35), the quantized energies E_n of, respectively, the symmetric and anti-symmetric bound states of the quantum well. The asymptotes of the tangent-like and minus-cotangent-like curves representing the left sides of Eqs. (4.37) and (4.40), respectively, in Figure 4.7 correspond exactly to the quantized $k_{2n} d$ values, $n\pi$, given in (4.25) for the infinite potential well case. Note that in the limit where V_0 or the radius of the circle corresponding to Eq. (4.38) becomes infinitely large, the crossing points are exactly at $k_2 d = n\pi$, as expected from the solutions of the infinite potential well. On the other hand, the crossing points in Figure 4.7 will always have values of $k_2 d$ somewhat smaller than $n\pi$. Thus, the wavelengths of the corresponding

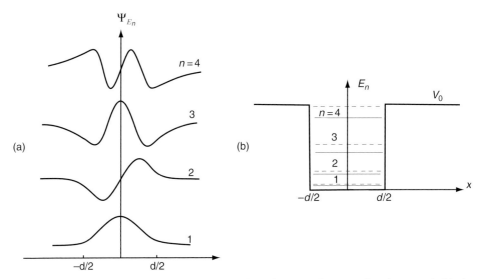

Figure 4.8. Schematic of (a) the symmetric and anti-symmetric wave functions, and (b) the quantized energies (solid lines) of the bound states of the finite potential well. The dashed lines show the corresponding energies in the limit $V_0 \to \infty$. (The horizontal axis in (a) is shifted for clarity.)

de Broglie waves of the bound states will always be longer in the finite potential well case than those in the corresponding infinite potential well case. The corresponding wave functions for the finite potential well case will not vanish at the boundaries at $x = d/2$ and $-d/2$ (see Figure 4.8a), allowing the wave functions in the wall regions I and III of the finite potential well to be finite. Since the wave functions are finite damped exponential waves in Regions I and III, they imply that the particle actually has a finite probability of penetrating the wall regions (see Figure 4.2) where the kinetic energy of the particle in the bound states is negative. Compared to the infinite potential well case, the over-all wave functions in the finite potential well case are more spread out than the corresponding wave functions in the infinite potential well case. According to Heisenberg's uncertainty principle, because the wave functions are more spread out, the uncertainty in the momentum, or the difference between the $+$ and $-|k_{2n}|$ values, must be smaller in the finite potential well case, as indeed is the case from Figure 4.7. The fact that the $|k_{2n}|$ values are smaller also means that the quantized energies are always down-shifted in the finite potential well case from the corresponding quantized energies in the infinite potential well (see Figure 4.8(b)). Furthermore, the closer is the quantized level to the top of the well, the larger is the down-shift, which is consistent with the fact the wave functions for these state are spread out more.

Finally, the number of bound states, N, can be determined easily from where the quarter-circle (4.48) crosses the horizontal axis $\alpha d = 0$ in the graph shown in Figure 4.7. The radius of the circle is $\sqrt{2mV_0d^2/\hbar^2}$; therefore, the condition $(N-1)\pi < \sqrt{2mV_0d^2/\hbar^2} < N\pi$ determines the value of N and, hence, the number of crossing points. For example, in Figure 4.7, the crossing point x on the horizontal axis is between 3π and 4π, and there are, indeed, four crossing points. If N is even, there is always the same number of symmetric and anti-symmetric states. If N is odd, there is always one more symmetric state than there are anti-symmetric states. Finally, no matter how small the radius of the circle is, or how narrow the well width is, there is always at least one symmetric bound state.

4.5 Three-dimensional potential box or quantum well

All the results given in the previous section can be easily extended to the case of a three-dimensional box where, for example:

$$
V(x, y, z) = \begin{cases} 0, & |x| < \dfrac{a}{2} \,,\ |y| < \dfrac{b}{2} \,,\ |z| < \dfrac{c}{2} \\[2mm] V_0, & |x| > \dfrac{a}{2} \,,\ |y| > \dfrac{b}{2} \,,\ |z| > \dfrac{c}{2}. \end{cases} \tag{4.41}
$$

The corresponding time-independent Schrödinger equation:

$$
\left[-\frac{\hbar^2}{2m} \left(\frac{\partial^2}{\partial x^2} + \frac{\partial^2}{\partial y^2} + \frac{\partial^2}{\partial z^2} \right) + V(x, y, z) \right] \Psi_{E_n}(x, y, z) = E_n \Psi_{E_n}(x, y, z), \tag{4.42}
$$

for a box with impenetrable walls at $x = \pm\dfrac{a}{2}$, $y = \pm\dfrac{b}{2}$, and $z = \pm\dfrac{c}{2}$, or $V_0 \to \infty$, can be solved by the standard method of separation of variables. Thus, one looks for particular solutions that can be factored in the following form:

$$
\Psi_{E_n}(x, y, z) = \Psi_{n_x}(x)\, \Psi_{n_y}(y)\, \Psi_{n_z}(z). \tag{4.43}
$$

Each factor in (4.43) can be found as a solution of a one-dimensional infinite potential well problem of the form considered in Section 4.4. The corresponding energy eigen values of the 3-D time-independent Schrödinger equation (4.42) is:

$$
E_{n_x n_y n_z} = E_{n_x} + E_{n_y} + E_{n_z}. \tag{4.44}
$$

Each quantized energy level is then specified by a set of three quantum numbers n_x, n_y, and n_z, where n_x, n_y, n_z each $= 1, 2, 3, \ldots$

In the multi-dimensional case, it is possible that several quantum states have the same energy value:

$$
E_{n_x n_y n_z} = E_{n'_x n'_y n'_z}.
$$

Such an energy level is said to be "*degenerate.*" The "degeneracy" of that level is equal to the number of states that have the same energy eigen value. For example, suppose the 3-D potential box has impenetrable walls ($V_0 \to \infty$) and the dimensions of the box are such that $a = b = 2c$. The energy eigen value of the box is then:

$$E_{n_x n_y n_z} = \frac{\pi^2 \hbar^2}{2m} \left[\frac{n_x^2}{a^2} + \frac{n_y^2}{b^2} + \frac{n_z^2}{c^2} \right]$$

$$= \frac{\pi^2 \hbar^2}{2ma^2} \left[n_x^2 + n_y^2 + 4n_z^2 \right].$$

(4.45)

The states (n_x, n_y, n_z) and (n_x', n_y', n_z') that satisfy the condition:

$$n_x^2 + n_y^2 + 4n_z^2 = n_x'^2 + n_y'^2 + 4n_z'^2$$

are degenerate. For example, the states:

n_x	n_y	n_z
1	2	2
1	4	1
2	1	2
4	1	1

are degenerate. The energy of the degenerate level is:

$$E_{n_x n_y n_z} = \frac{21\pi^2 \hbar^2}{2ma^2}$$

with a four-fold degeneracy.

4.6 Problems

4.1 Verify the expressions (4.20a) and (4.20b) given in the text for the transmission coefficient of a potential-barrier of height V_0 and width d. Derive the expression for T for the special case where $E = V_0$. Plot T versus (E/V_0), from (E/V_0) to 10, for $\beta = 2$ and $\beta = 6$, where $\beta \equiv [(2mV_0d^2)/\hbar^2]^{1/2}$ Assume $V_I = 0$ and $V_{II} = \Delta V$.

4.2 A particle with energy E in a region of zero potential is incident on a potential well of depth V_0 and width "d". From the expression for the probability of transmission T of the particle past the well given in (4.20a), find the approximate values of E (in terms of $\hbar^2/2md^2$) corresponding to the maxima and minima in T for (a) $\beta = 10$; (b) $\beta = 250$.

4.3 Consider a one-dimensional rectangular potential well structure such as that shown in Figure 4.9.

$$
\begin{aligned}
V &= V_1 && \text{for} && x < -a, \\
V &= 0 && \text{for} && -a < x < 0, \\
V &= V_1/2 && \text{for} && 0 < x < a, \\
V &= V_1 && \text{for} && x > a,
\end{aligned}
$$

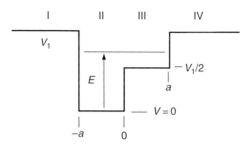

Figure 4.9. Multiple quantum well potential profile for problem 4.3.

Write the wave functions in regions I through IV and the equations (but do not try to solve these equations) describing the boundary conditions on these wave functions for

(a) $E > V_1$,
(b) $V_1 > E > V_1/2$,
(c) $E < V_1/2$.

4.4 Suppose the following wave function describes the state of an electron in an infinite square potential well, $0 < x < a$, with $V(x) = 0$ inside the well:

$$V(x) = \begin{cases} A \sin\left(\dfrac{3\pi x}{2a}\right) \cos\left(\dfrac{\pi x}{2a}\right) & \text{for} \quad 0 \le x \le a, \\ 0 & \text{elsewhere.} \end{cases}$$

(a) Normalize the wave function.
(b) Write down the full space- and time-dependent wave function $\Psi(x, t)$ that describe the state of the electron for all time.
(c) If measurements of the energy of the electron were made, what values of energy would be observed and with what absolute probabilities?

4.5 Consider the one-dimensional potential of Figure 4.10:

$$\begin{array}{lll} & & \text{Region} \\ V = \infty, & x < 0, & \text{I} \\ V = 0, & 0 < x < a, & \text{II} \\ V = V_0, & a > x. & \text{III} \end{array}$$

(a) Obtain, for this potential, the equation whose solution gives the eigen energies of the bound states ($E < V_0$).
(b) Sketch the eigenfunctions of the three lowest energies assuming V_0 is sufficiently large so that there are at least three bound states.

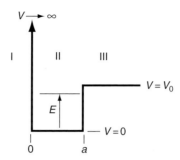

Figure 4.10. Quantum well potential profile for problem 4.5.

4.6 Consider the case of an electron ($m_e = 0.91 \times 10^{-27}$ g) in a finite potential well of depth 1.25 V and width 145 Å.

(a) First, estimate the number of bound states.
(b) Calculate the energies of the lowest two bound states.
(c) Sketch the wave functions for the lowest three bound states found in (b).

4.7 A particle of mass m is confined to move in a quantum well in the (x, y) plane which consists of a pair of impenetrable walls at $x = \pm a$ but is unbounded for motion in the y direction.

(a) Let the total energy of the particle be E and the energy associated with the motion in the x and y directions be E_x and E_y, respectively. What are the allowed values of E_x, E_y, and E?
(b) Sketch E versus k_y for various allowed values of E_x.
(c) Suppose the particle motion in the x direction corresponds to the second bound state of the infinite potential well and the total energy of the particle is E. Find the energy of the particle associated with its motion in the y direction.
(d) Find an acceptable, un-normalized, space- and time-dependent wave function to describe the particle in (c).
(e) If the particle's total energy is $E = \pi^2\hbar^2/4ma^2$, find the space- and time-dependent unnormalized wave function for the particle.
(f) Suppose now an infinite potential barrier at $y = \pm a$ is imposed. Can the particle's energy be measured to be $3\pi^2\hbar^2/4ma^2$? Why?

5 The harmonic oscillator and photons

The harmonic oscillator is a model for many physical systems of scientific and technological importance. It describes the motion of a bound particle in a potential well that increases quadratically with the distance from the minimum or the bottom of the potential well. Quantum mechanically, Heisenberg's equation of motion for the position of such a particle is of the same form as that of a classical harmonic oscillator. As such, it is a model for any physical system whose natural motion is described by the harmonic oscillator equation, such as the vibrational motion of molecules, lattice vibrations of crystals, the electric and magnetic fields of electromagnetic waves, etc. Quantization of the electromagnetic waves leads to the concepts of photons and coherent optical states. The eigen functions and quantized energies of harmonic oscillators in general share some general features with those of the square well potential considered in the previous chapter.

5.1 The harmonic oscillator based on Heisenberg's formalism of quantum mechanics

Consider the case of a point mass, m, attached to the end of a linear spring with a spring constant k (Figure 5.1). The classical equation of motion of the particle is:

$$m\frac{d^2x}{dt^2} = F_x(x) = -kx, \tag{5.1}$$

where x is the deviation of the position of the mass point from its equilibrium position. It can be put in the form of the harmonic oscillator equation:

$$\frac{d^2x}{dt^2} + \omega_0^2 x = 0, \tag{5.2}$$

where $\omega_0 = \sqrt{k/m}$ is the angular frequency of the oscillator.

The potential energy of the particle is (see Figure 5.1), from $-\frac{\partial}{\partial x}V(x) = F_x(x)$:

$$V(x) = -\int_o^x F_x(x')dx' = \frac{k}{2}x^2. \tag{5.3}$$

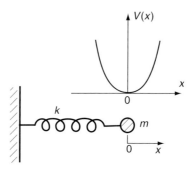

Figure 5.1. Schematic of a harmonic oscillator represented by a mass point attached to a linear spring. The potential energy of the particle varies quadratically with the deviation from its equilibrium position ($x = 0$).

The corresponding time-independent Schrödinger equation for the harmonic oscillator is then:

$$\hat{H}\Psi_{E_n} = \left[-\frac{\hbar^2}{2m}\frac{\partial^2}{\partial x^2} + \frac{k}{2}x^2 \right]\Psi_{E_n} = E_n\Psi_{E_n}. \tag{5.4}$$

Before attempting to solve Eq. (5.4), which is a differential equation with a variable coefficient and can be quite tedious to solve, it is instructive to derive Heisenberg's equation of motion for the harmonic oscillator according to Eq. (2.49) of Chapter 2 and compare it with the classical harmonic oscillator equation, (5.2). Thus,

$$\begin{aligned}\frac{d\hat{x}_t}{dt} &= \frac{i}{\hbar}\left[\hat{H}, \hat{x}_t \right] \\ &= \frac{i}{\hbar}\left[\left(\frac{\hat{p}_{xt}^2}{2m} + \frac{k}{2}\hat{x}_t^2 \right),\ \hat{x}_t \right].\end{aligned} \tag{5.5}$$

Making use of the commutation relationship (2.11a), we have:

$$\frac{d\hat{x}_t}{dt} = \frac{\hat{p}_{xt}}{m}. \tag{5.6a}$$

Similarly,

$$\begin{aligned}\frac{d\hat{p}_{xt}}{dt} &= \frac{i}{\hbar}\left[\hat{H}, \hat{p}_{xt} \right] \\ &= \frac{i}{\hbar}\left[\left(\frac{\hat{p}_{xt}^2}{2m} + \frac{k}{2}\hat{x}_t^2 \right),\ \hat{p}_{xt} \right] \\ &= -k\hat{x}_t.\end{aligned} \tag{5.6b}$$

Combining Eqs. (5.6a) and (5.6b) leads to the Heisenberg's equation of motion for the harmonic oscillator:

$$\frac{d^2 x_t}{dt^2} + \omega_0^2 x_t = 0, \tag{5.7}$$

which is of the same form as the classical equation of motion for the harmonic oscillator, (5.2). Thus, any dynamic variable that satisfies a classical equation of motion of the form (5.2) can be dealt with quantum mechanically as a harmonic oscillator, such as the vibrational motion of the normal mode coordinates of molecules.

Another very important class of examples is the radiation oscillators associated with electromagnetic waves. For instance, the electric field, $\varepsilon(z, t) = \varepsilon \cos(\omega t - kz - \phi)$, associated with a transverse electromagnetic wave at a given wavelength (or propagation constant $k = 2\pi/\lambda$) in free space satisfies the wave equation:

$$\frac{\partial^2 \varepsilon(z, t)}{\partial t^2} + k^2 c^2 \varepsilon(z, t) = 0. \tag{5.8}$$

In analogy with the mechanical harmonic oscillator, the radiation oscillators corresponding to the normal modes of the electromagnetic waves will, thus, also have particle properties in the form of photons. This point will be discussed in considerable detail in Section 5.4.

We will now derive the remarkable result that the energy eigen values, E_n, of the Hamiltonian of any harmonic oscillator are quantized and of the form:

$$E_n = (n + \frac{1}{2})\hbar\omega_o,$$

where n is an integer 0, 1, 2, 3, ... This can be done by solving the time-independent Schrödinger equation (5.4) to find the eigen values E_n, or by a clever use of the commutation relationship between $\hat{x}$ and $\hat{p}_x$, as shown by Dirac (see Dirac (1947), pp. 136–138). The former approach follows the standard procedure of using the method of power series expansion to solve ordinary differential equations with variable coefficients. It is tedious mathematically and does not offer a great deal of insights. We will postpone doing so for now. The latter approach is an interesting example of the power and elegance of using the operator relationships to solve quantum mechanical problems.

Suppose we introduce a pair of new variables $\hat{a}^+$ and $\hat{a}^-$ formed from the canonical variables $\hat{x}$ and $\hat{p}_x$:

$$\hat{a}^+ \equiv \frac{1}{\sqrt{2m\hbar\omega_0}}[\hat{p}_x + i\omega_0 m\hat{x}], \tag{5.9}$$

$$\hat{a}^- \equiv \frac{1}{\sqrt{2m\hbar\omega_0}}[\hat{p}_x - i\omega_0 m\hat{x}]. \tag{5.10}$$

Substituting these into the expression for the Hamiltonian, $\hat{H} = \frac{\hat{p}_x^2}{2m} + \frac{m\omega_0^2}{2}\hat{x}^2$, and making use of the commutation relationship (2.11a), we have:

$$\hat{H} = \left[\hat{a}^+\hat{a}^- + \frac{1}{2}\right]\hbar\omega_0, \tag{5.11}$$

and the commutation relationships:

$$[\hat{a}^-, \hat{a}^+] = 1, \quad [\hat{a}^-, \hat{a}^-] = 0, \quad [\hat{a}^+, \hat{a}^+] = 0. \tag{5.12}$$

To economize on notations, we now make use of Dirac's compact notations and let H' and $|H'\rangle$ be the eigen value and the corresponding eigen function of the Hamiltonian (5.11):

$$\hat{H}|H'\rangle = H'|H'\rangle. \tag{5.13}$$

It follows from (5.11) that:

$$\hbar\omega_0 \langle H'|\hat{a}^+\hat{a}^-|H'\rangle = \langle H'|(\hat{H} - \frac{1}{2}\hbar\omega_o)|H'\rangle$$

$$= (H' - \frac{1}{2}\hbar\omega_o)\langle H'|H'\rangle. \tag{5.14}$$

Since the magnitude of the ket vector $\hat{a}^-|H'\rangle$ is always greater than or equal to zero,

$$|(\hat{a}^-|H'\rangle)|^2 = \langle H'|\hat{a}^+\hat{a}^-|H'\rangle \geq 0.$$

The equality occurs only for $|(\hat{a}^-|H'\rangle)| = 0$. Thus, for a non-trivial eigen state, $\langle H'|H'\rangle \neq 0$, it follows from (5.14) that:

$$H' \geq \frac{1}{2}\hbar\omega_0, \tag{5.15}$$

and the minimum energy, E_0, of the harmonic oscillator is not zero but:

$$E_0 = \frac{1}{2}\hbar\omega_0; \tag{5.16}$$

this occurs when and only when $\hat{a}^-|H'\rangle = 0$. Let the corresponding *normalized* eigen state $|H'\rangle$ belonging to this eigen value be designated $|0\rangle$ thus:

$$\hat{H}|0\rangle = \frac{1}{2}\hbar\omega_0|0\rangle, \tag{5.17}$$

$$\hat{a}^-|0\rangle = 0, \tag{5.18}$$

and

$$\langle 0|0\rangle = 1. \tag{5.19}$$

Starting with (5.17) and making use of the commutation relationship (5.12), it can be shown that $\hat{a}^+|0\rangle$ is also an eigen state of $\hat{H}$, but the corresponding eigen value is $\left(1 + \frac{1}{2}\right)\hbar\omega_0$, or:

$$\hat{H}\hat{a}^+|0\rangle = \left(1 + \frac{1}{2}\right)\hbar\omega_0 \hat{a}^+|0\rangle. \tag{5.20}$$

Through repeated use of this procedure, it can be shown that $\hat{a}^+(\hat{a}^+|0\rangle)$, $\hat{a}^+(\hat{a}^+\hat{a}^+|0\rangle)$, etc. are also eigen states of the Hamiltonian corresponding to the eigen values $(2+1/2)\hbar\omega_0$, $(3+1/2)\hbar\omega_0, \ldots$, respectively. Finally, we have the remarkable conclusion that the eigen values of the Hamiltonian, or the energy of the harmonic oscillator, is quantized in units of $\hbar\omega_0$:

$$H' = E_n = \left(n + \frac{1}{2}\right)\hbar\omega_0, \tag{5.21}$$

where n is an integer 0, 1, 2, $\ldots$, and the lowest energy state is not zero but $\hbar\omega_0/2$. The corresponding eigen states are: $(\hat{a}^+)^n|0\rangle$.

Since the derivation of these results does not depend in any way on the physical nature of the oscillator, be it mechanical or electrical, the energy of the electromagnetic radiation oscillators is also expected to be quantized in units of $\hbar\omega_0$; therefore, a light wave can be viewed as consisting of particles, or photons, of energy $\hbar\omega_0$. The fact that the minimum energy is not equal to zero is expected on the basis of Heisenberg's uncertainty principle, just as in the square well potential case considered in Chapter 4, because the particle is localized in the well. This energy is called the "zero-point energy" of the oscillator, representing the energy due to the fluctuating motions of the particle around its equilibrium position at $x=0$ in the ground state of the oscillator. In the harmonic oscillator case, the position of the particle cannot be localized exactly at $x=0$; for, otherwise, the kinetic energy will have to be infinitely large according to the uncertainty principle. On the other hand, the kinetic energy also cannot be zero. For if it were so, the wave function will have infinite width and the potential energy would have to be infinitely large. Thus, the total energy in the minimum energy state must be partly potential energy and partly kinetic energy. In fact, it can be shown that the expectation values of the kinetic energy and the potential energy in any eigen state of the harmonic oscillator are always equal. This is also a result of what is called the virial theorem. In the case of electromagnetic waves, the zero-point energy corresponds to the fluctuations of the electric and magnetic fields in vacuum, or the "vacuum fluctuations" of the radiation fields.

Returning now to the eigen states of the Hamiltonian, although the states $(\hat{a}^+)^n|0\rangle$ are eigen states of the Hamiltonian belonging to the eigen value $\left(n + \frac{1}{2}\right)\hbar\omega_0$, for $n = 0, 1, 2, \ldots$, they are not necessarily *normalized* eigen states. Let us designate the *normalized* eigen states $|0\rangle, |1\rangle, |2\rangle, \ldots |n\rangle, \ldots$, so that:

$$\langle 0|0\rangle = \langle 1|1\rangle = \langle 2|2\rangle = \cdots = \langle n|n\rangle = 1. \tag{5.22}$$

These normalized eigen states are state by state proportional to the $(\hat{a}^+)^n|0\rangle$ states. The proportionality constants for different states are different, and the state $|n\rangle$ is not equal to, but only proportional to $\hat{a}^+|n-1\rangle$, for example. Let

$$\hat{a}^+|n-1\rangle = \alpha_n|n\rangle,$$

where the proportionality constant α_n is a real number. Thus, using the commutation relationship (5.12) and since both $|n\rangle$ and $|n-1\rangle$ are normalized:

$$\langle n - 1|\hat{a}^-\hat{a}^+|n - 1\rangle = \langle n - 1|(1 + \hat{a}^+\hat{a}^-)|n - 1\rangle$$
$$= [1 + (n - 1)]\langle n - 1|n - 1\rangle = n;$$

at the same time,

$$\langle n - 1|\hat{a}^-\hat{a}^+|n - 1\rangle = \langle n|\,\alpha_n\alpha_n|n\rangle = (\alpha_n)^2\langle n|n\rangle = (\alpha_n)^2.$$

Therefore,

$$\alpha_n = \sqrt{n},$$

and

$$\hat{a}^+|n - 1\rangle = \sqrt{n}|n\rangle. \tag{5.23a}$$

Similarly, it can be shown that:

$$\hat{a}^-|n\rangle = \sqrt{n}|n - 1\rangle, \tag{5.23b}$$

and

$$\langle n|\hat{a}^+\hat{a}^-|n\rangle = n. \tag{5.24}$$

Therefore, *a complete orthonormal set of eigen functions* for the quantized energy states can be generated from the ground state wave function of the harmonic oscillator through repeated use of the operator $\hat{a}^+$:

$$|1\rangle = \frac{\hat{a}^+}{\sqrt{1}}|0\rangle$$

$$|2\rangle = \frac{\hat{a}^+}{\sqrt{2}}|1\rangle = \frac{(\hat{a}^+)^2}{\sqrt{2!}}|0\rangle$$

$$|3\rangle = \frac{\hat{a}^+}{\sqrt{3}}|2\rangle = \frac{(\hat{a}^+)^3}{\sqrt{3!}}|0\rangle$$

.

.

.

$$|n\rangle = \frac{\hat{a}^+}{\sqrt{n}}|n - 1\rangle = \frac{(\hat{a}^+)^n}{\sqrt{n!}}|0\rangle, \tag{5.25}$$

and all these $|n\rangle$ states are now normalized as in (5.22).

Since $|n - 1\rangle$ and $|n\rangle$ represent quantum states of the harmonic oscillator with $n - 1$ and n quanta of energy $\hbar\omega_0$, respectively, (5.23a) and (5.25) show clearly that the effect of the operator $\hat{a}^+$ on the oscillator state with $n-1$ quanta is to create an additional quantum and change it to the state with n quanta. $\hat{a}^+$, therefore, has the meaning of a "creation operator." Similarly, the effect of the operator $\hat{a}^-$ on the oscillator state with n quanta is to decrease the number of quanta from n to $n-1$. $\hat{a}^-$, therefore, has the meaning of an "annihilation operator." Furthermore, in the context of the discussions of Heisenberg's matrix formulation of quantum mechanics in Section 2.5, the matrix representation of the creation operator $\hat{a}^+$ using the eigen states $|n\rangle$ as the basis states is, from (5.23a): $\langle n|\hat{a}^+|n'\rangle = \sqrt{n}\,\delta_{n',n-1}$, and the annihilation operator is, from (5.23b): $\langle n'|\hat{a}^-|n\rangle = \sqrt{n-1}\,\delta_{n',n-1}$, or:

$$\hat{a}^+ = \begin{pmatrix} 0 & 0 & 0 & 0 & 0 & 0 & \cdot & \cdot & \cdot \\ \sqrt{1} & 0 & 0 & 0 & 0 & 0 & \cdot & \cdot & \cdot \\ 0 & \sqrt{2} & 0 & 0 & 0 & 0 & \cdot & \cdot & \cdot \\ 0 & 0 & \sqrt{3} & 0 & 0 & 0 & \cdot & \cdot & \cdot \\ \cdots & \cdots & \cdots & & & & \end{pmatrix}, \text{ and}$$

$$\hat{a}^- = \begin{pmatrix} 0 & \sqrt{1} & 0 & 0 & 0 & 0 & \cdot & \cdot & \cdot \\ 0 & 0 & \sqrt{2} & 0 & 0 & 0 & \cdot & \cdot & \cdot \\ 0 & 0 & 0 & \sqrt{3} & 0 & 0 & \cdot & \cdot & \cdot \\ 0 & 0 & 0 & 0 & \sqrt{4} & 0 & \cdot & \cdot & \cdot \\ \cdots & \cdots & \cdots & & & & \end{pmatrix}. \tag{5.26}$$

The matrix representation of the Hamiltonian in the same basis is diagonal, and from (5.24): $\langle n|\hat{H}|n'\rangle = \left(n+\frac{1}{2}\right)\hbar\omega_0\,\delta_{n,n'}$, and its diagonal elements are the quantized energies of the harmonic oscillator:

$$\hat{H} = \begin{pmatrix} \frac{1}{2} & 0 & 0 & 0 & 0 & 0 & \cdot & \cdot & \cdot \\ 0 & \frac{3}{2} & 0 & 0 & 0 & 0 & \cdot & \cdot & \cdot \\ 0 & 0 & \frac{5}{2} & 0 & 0 & 0 & \cdot & \cdot & \cdot \\ 0 & 0 & 0 & \frac{7}{2} & 0 & 0 & \cdot & \cdot & \cdot \\ \cdots & \cdots & \cdots & \cdots & & & \end{pmatrix} \hbar\omega_0. \tag{5.27}$$

The eigen states in the matrix representation using these states as the basis states are simply:

$$|0\rangle = \begin{pmatrix} 1 \\ 0 \\ 0 \\ 0 \\ \vdots \end{pmatrix}, \quad |1\rangle = \begin{pmatrix} 0 \\ 1 \\ 0 \\ 0 \\ \vdots \end{pmatrix}, \quad |2\rangle = \begin{pmatrix} 0 \\ 0 \\ 1 \\ 0 \\ \vdots \end{pmatrix}, \quad |3\rangle = \begin{pmatrix} 0 \\ 0 \\ 0 \\ 1 \\ \vdots \end{pmatrix}, \text{etc.} \tag{5.28}$$

With the matrix representations, one can easily verify that all the operator relationships and equations, from (5.12) to (5.25), for the harmonic oscillator are satisfied. Thus, with the exception of the wave functions of the quantized energy states in the Schrödinger representation, all the quantum mechanical properties of the harmonic oscillator are now formally known.

The conventional approach to derive the wave functions of the energy eigen states in Schrödinger's representation is to solve the time-independent Schrödinger equation, (5.4), for the harmonic oscillator. However, with the results already obtained in this section, (5.23a) – (5.25), it is possible to obtain these wave functions in a uniquely simple way, which does not require ever having to solve any differential equation with variable coefficients in a complicated way. It is based on Eqs. (5.18), (5.19), and (5.25). Using the annihilation operator defined in (5.10), Eq. (5.18) in the Schrödinger representation is:

$$\hat{a}^- \langle x|0\rangle = \frac{1}{\sqrt{2m\hbar\omega_0}} \left[-i\hbar \frac{\partial}{\partial x} - im\omega_0 x\right] \Psi_{E_0}(x) = 0. \tag{5.29}$$

Introducing the variable $z = x^2$, (5.29) becomes:

$$\frac{\partial}{\partial z} \Psi_{E_0}(z) = -\frac{m\omega_0}{2\hbar} \Psi_{E_0}(z);$$

therefore,

$$\Psi_{E_0}(z) = C_0 e^{-\frac{m\omega_0}{2\hbar} z}, \tag{5.30}$$

where C_0 is a constant to be determined by the normalization condition. Converting z back to x^2 and applying the normalization condition (5.19), we obtain the ground-state wave function of the harmonic oscillator:

$$\Psi_{E_0}(x) = \left(\frac{m\omega_0}{\hbar\pi}\right)^{1/4} e^{-\frac{m\omega_0}{2\hbar} x^2}. \tag{5.31}$$

Note that this wave function of the ground state of the harmonic oscillator is a Gaussian wave packet. As such, it is a "minimum-uncertainty wave packet" as the discussion following (3.25) showed.

Using the Schrödinger representation of the creation operator defined in (5.9), one can then generate, without solving any differential equations, the wave functions for all the other quantized states of the harmonic oscillator according to (5.25). With these and earlier results, all the formal properties of the harmonic oscillator are now known. We will discuss their physical significance after the discussion on the use of the alternative Schrödinger approach to solve the problem in the following section.

5.2 The harmonic oscillator based on Schrödinger's formalism

The conventional way to deal with the harmonic oscillator problem is to find the eigen values and eigen functions of the Hamiltonian by solving the time-independent

Schrödinger equation in the form (5.4). The standard method for solving such a differential equation with a variable coefficient is to expand the solution in a power series of the independent variable x and then find the recursion relationships for all the expansion coefficients. Many such equations have, however, already been solved in the past and the solutions are well known and tabulated. Therefore, there is no need to barge ahead to try to solve (5.4) from scratch by the same method. It behooves us to see first whether the equation can be recast in a form that fits one of these equations that have already been solved years ago. It turns out that indeed it can be. By making the following substitution:

$$\Psi_{E_n}(x) = C_n H_n\left(\sqrt{\frac{m\omega_0}{\hbar}}x\right)e^{-\frac{m\omega_0}{2\hbar}x^2}, \tag{5.32}$$

Eq. (5.4) for $\Psi_n(x)$ can be transformed into one for $H_n(u)$:

$$\frac{d^2 H_n(u)}{du^2} - 2u\frac{dH_n(u)}{du} + 2\left(\frac{E_n}{\hbar\omega_0} - \frac{1}{2}\right)H_n(u) = 0, \tag{5.33}$$

where $u = \left(\sqrt{\frac{m\omega_0}{\hbar}}\right)x$. The solutions of Eq. (5.33) that correspond to the bounded states of the harmonic oscillator ($|\Psi_{E_n}(x)|$ does not diverge as $x \to \pm\infty$) are well known as the Hermite polynomials:

$$H_0(u) = 1,$$
$$H_1(u) = 2u,$$
$$H_2(u) = -2 + 4u^2,$$
$$H_3(u) = -12u + 8u^3,$$
$$H_4(u) = 12 - 48u^2 + 16u^4,$$
$$\vdots$$
$$H_n(u) = (-1)^n e^{u^2}\frac{d^n}{du^n}e^{-u^2},$$
$$\vdots \tag{5.34}$$

which satisfy the recursion relationship:

$$\frac{dH_n(u)}{du} = 2nH_{n-1}(u). \tag{5.35}$$

The corresponding eigen values are:

$$\frac{E_n}{\hbar\omega_0} - \frac{1}{2} = n = 0, 1, 2, 3, 4, \cdots \quad \text{or} \quad E_n = \left(n + \frac{1}{2}\right)\hbar\omega_0. \tag{5.36}$$

The normalization constant is:

$$C_n = \frac{1}{\sqrt{2^n n!}}\left(\frac{m\omega_0}{\hbar\pi}\right)^{1/4}. \tag{5.37}$$

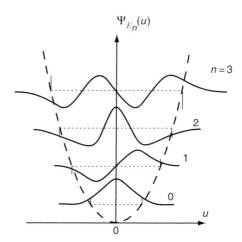

Figure 5.2. Schematic of the wave functions for the harmonic oscillator. The dashed line corresponds to $V(u)$ and the dotted line corresponds to the quantized energy levels of the oscillator. ($u \equiv \sqrt{m\omega_0/\hbar}\, x$.)

These results are exactly the same as those obtained in the previous section, 5.1: (5.32), (5.34), and (5.37) are equivalent to (5.25) and (5.31); (5.36) is exactly the same as (5.21).

The first four wave functions are shown schematically in Figure 5.2 as an illustration. Qualitatively, they are similar to the wave functions found in Section 4.4 for the finite square well potential and can be understood on the same basis. First of all, the potential of the harmonic oscillator is symmetric under inversion of the coordinate axis ($x \rightarrow -x$); therefore, the parity of the wave functions is well defined. The wave functions are either symmetric ($n = 0, 2, 4, \ldots$) or anti-symmetric ($n = 1, 3, 5, \ldots$) with respect to x. Second, within the well-width where the kinetic energy of the oscillator is positive, the wave functions consist of de Broglie waves of spatially-varying speeds propagating in opposite directions, giving rise to the sinusoidal-like interference patterns. The energy is totally kinetic energy in the middle of the well and totally potential energy at the edges of the well. There is an exponentially smaller probability of finding the particle penetrating the regions outside the well width where the kinetic energy is negative and the wave functions are decaying functions.

The result on the energy eigen values of the harmonic oscillator, (5.21), is of fundamental importance. It has two key points. First, the energy of the harmonic oscillator is quantized; second, it is quantized in units of $\hbar\omega_0$. As pointed out earlier, the harmonic oscillator is also a model for the radiation oscillators of the electromagnetic wave. The fact that the electromagnetic wave is quantized was a totally new concept before the discovery of the principles of quantum mechanics and quantum electrodynamics. In fact, as will be shown in Section 5.4 below, it was the attempt to solve the black-body radiation problem that led Planck to his far-reaching pioneering hypothesis that the energy of the radiation oscillators is quantized and in units of $\hbar\omega_0$. It was one of the key steps in the development of quantum mechanics.

The seminal idea that light waves are also particles of photons has led to the successful explanations of numerous physical phenomena and countless applications of great importance. Although Planck's postulate on the radiation oscillators can be considered a basic postulate of quantum mechanics, as we have seen from the discussions in this chapter, it is also a direct mathematical consequence of two even more fundamental postulates: Heisenberg's uncertainty principle which relates the momentum operators to the spatial derivative operators (Postulate 2, Section 2.2), and Schrödinger's equation which relates the time-derivative operator to the Hamiltonian or the energy (Postulate 3, Section 2.3) of all dynamic systems. These basic postulates are at the roots of every quantum phenomenon in the atomic and sub-atomic world and have now been verified experimentally again and again in countless experiments without exception, so far.

5.3   Superposition state and wave packet oscillation

The wave functions found in the previous sections are eigen functions of the Hamiltonian. As such they are stationary states, and the corresponding probability distribution functions are independent of time. How then can we reconcile the information obtained on the wave functions and the quantized energies with the simple classical picture of the natural oscillating motion of a mass point at the end of a spring bouncing back and forth around its equilibrium position at $x = 0$, as shown in Figure 5.1? To describe such an oscillating motion according to quantum mechanics, we must take the time-dependence of the wave functions into account and form a superposition state, $\Psi(x, t)$, of these wave functions to correspond to the state of the mass point that is localized at an initial non-equilibrium position, say, $x = x_1$. If we know the mass point is localized at this position, its state function $\Psi(x, 0)$ cannot be an eigen function of the Hamiltonian. It must be a superposition of the eigen functions, or a mixed state:

$$\Psi(x, 0) = \sum_n C_n \Psi_{E_n}(x, 0), \tag{5.38}$$

where $\sum_n |C_n|^2 = 1$. The precise form of $\Psi(x, 0)$ is not important for the present general discussion. It is sufficient to say that it is a sharply peaked wave packet centered on $x = x_1$. At a later time t, the state function becomes:

$$\Psi(x, t) = \sum_n C_n \Psi_{E_n}(x, t)$$
$$= \sum_n C_n \Psi_{E_n}(x) e^{-\frac{i}{\hbar}E_n t},$$

from (2.24). The eigen states have either even (e) or odd (o) parity, and $E_n = (n + 1/2)\hbar\omega_0$; therefore,

$$\Psi(x,t) = \Psi^{(e)}(x,t) + \Psi^{(o)}(x,t)$$

$$= e^{-i\frac{\omega_0 t}{2}} \left\{ \sum_{n=0,2,4,\cdots} C_n \Psi_{E_n}^{(e)}(x) e^{-in\omega_0 t} + \sum_{n=1,3,5,\cdots} C_n \Psi_{E_n}^{(o)}(x) e^{-in\omega_0 t} \right\}. \quad (5.39)$$

After an odd number of half-cycles, or $\omega_0 t = \pi$, 3π, $5\pi \ldots (2N+1)\pi$, $\ldots$, where N is an integer equal to $0, 1, 2, 3, \ldots$, it follows from (5.39) that the probability distribution function for the position of the mass point is:

$$|\Psi(x, t = (2N+1)\pi/\omega_o)|^2 = \left| \sum_{n=0,2,4,\ldots} C_n \Psi_{E_n}^{(e)}(-x) + \sum_{n=1,3,5,\ldots} C_n \Psi_{E_n}^{(o)}(-x) \right|^2$$

$$= |\Psi(-x, t = 0)|^2, \quad (5.40a)$$

which means the initial wave packet is reproduced exactly on the opposite side of the equilibrium point and centered on $x = -x_1$. Similarly, after an even number of half-cycles, or $\omega_0 t = 2\pi$, 4π, $6\pi, \ldots, 2N\pi, \ldots$, the corresponding probability distribution function becomes:

$$|\Psi(x, t = 2N\pi/\omega_o)|^2 = |\Psi(x, t = 0)|^2, \quad (5.40b)$$

or the initial wave packet is exactly reproduced in the original position centered on $x = x_1$. In summary, the initial wave packet oscillates at the angular frequency ω_0 back and forth between the extreme points at $x = \pm x_1$ according to:

$$|\Psi(x, t = N\pi/\omega_o)|^2 = |\Psi[(-1)^N x, t = 0]|^2, \quad (5.41)$$

where N is an integer $= 0, 1, 2, 3, \ldots$, similar to the classical picture of the motion of a harmonic oscillator. In between the extreme points the wave packet may disperse and change in shape somewhat.

For a better understanding of the formal mathematical proof of the wave packet oscillation phenomenon given above, let us limit the superposition state of the initial wave packet to two eigen states, one with even parity and one with odd parity:

$$\Psi(x, t = 0) = \Psi_{E_0}^{(e)}(x, 0) + \Psi_{E_1}^{(o)}(x, 0), \quad (5.42)$$

as shown in Figure 5.3(a). With just two states, the wave packet is, of course, not as localized as one that can be formed from the complete set of eigen functions. At half a cycle later, or $t = \pi/\omega_0$, the $n = 1$ component in (5.42) acquires a minus sign:

$$\Psi(x, t = \pi/\omega_0) = e^{-i\pi/2} \left\{ \Psi_{E_0}^{(e)}(x) + \Psi_{E_1}^{(o)}(x) e^{-i\pi} \right\}$$

$$= -i \left\{ \Psi_{E_0}^{(e)}(x) - \Psi_{E_1}^{(o)}(x) \right\}. \quad (5.43)$$

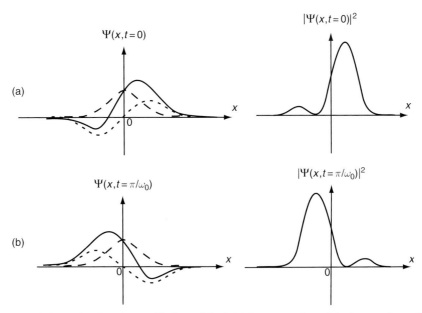

Figure 5.3. Schematics showing oscillation of the initial wave packet of the harmonic oscillator around the equilibrium point $x = 0$, (a) at $t = 0$, and (b) at $t = \pi/\omega_0$. Dashed curves: $\Psi_{E_0}^{(e)}(x)$. Dotted curves: $\Psi_{E_1}^{(o)}(x)$. Solid curves: $\Psi(x, t)$ and $|\Psi(x, t)|^2$ (arbitrary scale).

Mathematically, this odd-parity state $-\Psi_{E_1}^{(o)}(x)$ is equal to $\Psi_{E_1}^{(o)}(-x)$, or equivalently to the wave function resulting from inversion of the coordinate axis x. For the even-parity state, $\Psi_{E_0}^{(e)}(x)$ is the same as $\Psi_{E_0}^{(e)}(-x)$. Thus,

$$|\Psi(x, t = \pi/\omega_0)| = \left| -i\left\{ \Psi_{E_0}^{(e)}(-x) + \Psi_{E_1}^{(o)}(-x) \right\} \right| = |\Psi(-x, 0)|. \tag{5.44}$$

Physically, as clearly shown in Figure 5.3(b), the corresponding wave packet is now a reflection in $x(x \to -x)$ of the initial wave packet. Repeating this procedure for successive half cycles leads to oscillation of the wave packet back-and-forth around $x = 0$ and to a result for the two-state wave packet that is in complete agreement with Eq. (5.41).

5.4 Photons

Until now, we have been concentrating mainly on the wave nature of classical particles. In this section, we will explore the particle nature of classical waves, such as the duality of electromagnetic waves and photons on the basis of the analogy between harmonic oscillators and radiation oscillators, as mentioned briefly in Sections 5.1 and 5.2. The rules of quantization of electromagnetic waves leading to the concept of photons will be introduced. The validity of these rules and the concept of photons are confirmed by comparing the experimentally observable black-body radiation spectrum with Planck's quantum mechanical radiation law. The quantum mechanical coherent optic state which reduces to the classical coherent electromagnetic wave in the limit of large photon numbers is introduced.

Quantization of electromagnetic waves

Classical electromagnetic waves are characterized by a precisely measurable electric field $\vec{E}(\vec{r}, t)$ and a magnetic field $\vec{B}(\vec{r}, t)$ at every spatial point and every instant of time. These fields satisfy Maxwell's equations in free space:

$$\nabla \cdot \vec{D} = 0, \tag{5.45a}$$

$$\nabla \cdot \vec{B} = 0, \tag{5.45b}$$

$$\nabla \times \vec{E} = -\frac{1}{c}\frac{\partial \vec{B}}{\partial t}, \tag{5.45c}$$

$$\nabla \times \vec{H} = \frac{1}{c}\frac{\partial \vec{D}}{\partial t}, \tag{5.45d}$$

and can be expanded in a complete set of orthonormal "modes", $\vec{u}_k(\vec{r})$, corresponding to the frequency ω_k, that satisfy the equation:

$$\nabla^2 \vec{u}_k(\vec{r}) + \frac{\omega_k^2}{c^2}\vec{u}_k(\vec{r}) = 0 \tag{5.46a}$$

and the prescribed boundary conditions. For transverse electromagnetic waves, $\vec{u}_k(\vec{r})$ also satisfies the condition:

$$\nabla \cdot \vec{u}_k(\vec{r}) = 0. \tag{5.46b}$$

To simplify the discussion to follow, we assume a single-mode plane wave with the electric field linearly polarized in the x direction and propagating in the z direction in free space. The normalized mode function for a length $L \gg \lambda$ and unit area of free space is, therefore:

$$\vec{u}_k(\vec{r}) = L^{-1/2}e^{ikz}\, \mathbf{e}_x, \tag{5.47}$$

where $\mathbf{e}_x$ is a unit polarization vector and $\omega_k^2 = k^2 c^2$. The corresponding electric field can, thus, be written in the form:

$$\vec{E}(\vec{r}, t) \equiv E_x(z, t)\mathbf{e}_x = i\sqrt{\frac{2\pi\hbar\omega_k}{L}}\left(a_k^- e^{-i\omega_k t + ikz} - a_k^+ e^{i\omega_k t - ikz}\right)\mathbf{e}_x, \tag{5.48a}$$

where a_k^+ and $a_k^- \equiv (a_k^+)^*$ are ordinary variables proportional to the complex amplitude of the classic electric and magnetic fields of the particular plane-wave mode k.

The particular form of (5.48a) may appear to be somewhat arbitrary at this point. The proportionality constant $i\sqrt{2\pi\hbar\omega_k/L}$ is there so that the complex amplitudes $a_k^{\pm}$ so defined can be more easily compared with the corresponding parameters in the harmonic oscillator problem discussed in Section 5.1, as will be shown below. Substituting (5.48a) into (5.45c) gives:

$$\vec{B}(\vec{r}, t) \equiv B_y(z, t)\,\mathbf{e}_y = i\sqrt{\frac{2\pi\hbar\omega_k}{L}}\left(a_k^- e^{-i\omega_k t + ikz} - a_k^+ e^{i\omega_k t - ikz}\right)\mathbf{e}_y. \qquad (5.48b)$$

Because the electric and magnetic fields are precisely measurable properties of the classical electromagnetic wave, the magnitude and phase of the complex amplitude of the wave can be specified precisely simultaneously. As will be seen below, this is not the case for quantized fields.

Note that both $E_x(z, t)$ and $B_y(z, t)$ satisfy the wave equation of the form (5.8), which is analogous to the harmonic oscillator equation of the form (5.7). Based on this analogy with the harmonic oscillator, to quantize the fields, the dynamic variables corresponding to the complex amplitudes $a_k^{\pm}$ of the fields become operators $\hat{a}_k^{\pm}$ that satisfy the commutation rules:

$$[\hat{a}_k^-, \hat{a}_k^+] = 1, \quad [\hat{a}_k^-, \hat{a}_k^-] = 0, \quad [\hat{a}_k^+, \hat{a}_k^+] = 0. \qquad (5.49)$$

Like the basic postulates in quantum mechanics (as discussed in Chapter 2), there is no a priori reason to expect that these rules of quantization for electromagnetic waves would be correct. Their validity can only be established by comparing the predictions based upon these rules with experimental results. As will be shown in the discussion in the following subsection on the black-body radiation spectrum and Planck's radiation law, these rules are indeed correct. This conclusion is further verified in countless other experiments.

The classical electromagnetic energy in free space of unit cross-sectional area and length L is (in unrationalized cgs Gaussian units, $\varepsilon_0 = 1$ and $\mu_0 = 1$):

$$\frac{1}{8\pi}\int_0^L \left[E_x^2(z, t) + B_y^2(z, t)\right]dz.$$

From (5.48a & b) and the commutation rules (5.49), the corresponding Hamiltonian of the fields is, therefore:

$$\hat{H} = [\hat{a}_k^+ \hat{a}_k^- + \frac{1}{2}]\hbar\omega_k. \qquad (5.50)$$

The commutation rules, (5.49), and the Hamiltonian, (5.50), of the radiation oscillator are exactly the same as those of the harmonic oscillator, (5.12) and (5.11), respectively. Thus, all the results obtained for the harmonic oscillator obtained in Section 5.1–5.3 are directly applicable to the radiation oscillators. One of the most important results is that electromagnetic waves also have particle properties in the sense that the energy in each normal mode of the wave is quantized in units of $\hbar\omega$, or $h\nu$, with the energy eigen values of the Hamiltonian:

$$E_{kn} = \left(n_k + \frac{1}{2}\right)\hbar\omega_k, \text{ where } n_k = 0, 1, 2, 3, \ldots \qquad (5.50a)$$

and the corresponding eigen states are the fixed-photon-number states $|n_k\rangle$. n_k is the photon number per radiation mode in a volume of unit cross-sectional area and length L of the medium.

It is of interest to note that, because $\hat{a}_k^-$ and $\hat{a}_k^+$ do not commute, it implies that an uncertainty relationship exists between the intensity, or the photon number n, and the phase ϕ of the light wave. (See, for example, W. Heitler (1954), p. 65.) Qualitatively, since the time-dependence of the solutions of the Schrödinger equation is $e^{-\frac{iE_n t}{\hbar}}$, in analogy with the uncertainty relationship between the momentum and the coordinate variables of the harmonic oscillator, there is an uncertainty relationship between the variables E_n and t: $\Delta E_n \Delta t = \Delta(n\hbar)\Delta(\omega t) \geq (\sim\hbar)$, which leads to an uncertainty relationship between the photon number and phase of the light wave: $\Delta n \Delta \phi \geq (\sim 1)$. Thus, quantum mechanically, one cannot know the magnitude and the phase of the complex amplitude of the fields precisely simultaneously. In the case of a classical coherent single-mode monochromatic optical wave, the phase of the wave is known accurately, or $\Delta\phi \approx 0$. On the other hand, the uncertainty relationship implies that, quantum mechanically, the uncertainty in the photon number Δn must be large. Since in a classical wave the intensity can also be specified accurately or $\Delta n/\langle n\rangle \approx 0$, the expectation value of the photon number $\langle n\rangle$ in such a wave must then be very much larger than Δn. Thus, the classical description is good only when the photon number in the optical wave is large. For weak optical beams, the quantum description must be used. These subtle points will be discussed in more detail and made more quantitative later in this section in connection with the quantum theory of coherent optical states.

Black-body radiation

Historically, it was the attempt to resolve the puzzling obvious discrepancy between the classical theory of black-body radiation and the experimental observations that led Planck to postulate in the first place that light waves must also be particles in the form of photons. One simple test that confirmed the validity of the basic rules of quantization of electromagnetic waves (5.49) and the related results was that only quantum theory could correctly explain the "black-body radiation spectrum."

A "black-body" is a body that absorbs electromagnetic radiation completely at all wavelengths; thus it appears totally 'black'. A model of an ideal "black-body" is a completely enclosed cavity, like a light-proof dark room, at 0 K temperature. Looking in from outside through a tiny observation hole in the wall, the interior of the cavity will appear pitch dark because any light that gets into the cavity through the small hole will bounce back-and-forth all around the cavity and be absorbed by the cavity walls eventually with a very small probability of re-emitting from the tiny hole. Thus, it is totally black. If the cavity walls are at a finite temperature T, then the atoms in the wall will radiate heat in the form of electromagnetic radiation. In thermal equilibrium, the thermal radiation inside the cavity will be at equilibrium with the wall at the temperature T. A small amount of the thermal radiation can escape from the small hole as the "black-body radiation" and be measured by an external detector. What is the spectrum of the thermal radiation from this ideal black-body? We will first try to find the answer to this simple question on the basis of classical physics and see that the answer cannot possibly be correct.

The black-body radiation spectrum is a replica of the spectrum of the thermal energy spectrum inside the cavity. It can be determined from the thermal energy $\langle E_{th} \rangle$ per mode and the density-of-modes, $D(\nu) \equiv \Delta N / (V \cdot \Delta \nu)$, which is defined as the number of electromagnetic radiation modes per volume per frequency interval from ν to $\nu + \Delta \nu$.

Consider a cavity with linear dimensions very much larger than the wavelength in the wavelength range of interest, so that the radiation modes are essentially the same as those in free space. The shape of the cavity does not matter. For definiteness, let us assume it to be a cubic cavity of linear dimension L. The boundary conditions of the fields inside the cavity also do not matter for a large enough cavity. Assume the modes inside satisfy the periodic conditions and are of the form $e^{i\vec{k}\cdot\vec{r}}$, where

$$k_x = \pm \frac{2\pi N_x}{L}, k_y = \pm \frac{2\pi N_y}{L}, \text{ and } k_z = \pm \frac{2\pi N_z}{L}. \tag{5.51}$$

From Maxwell's equations:

$$\omega^2 = k^2 c^2 = (k_x^2 + k_y^2 + k_z^2)c^2. \tag{5.52}$$

Each set of (k_x, k_y, k_z) values corresponds to a possible propagation mode and each propagation mode has two polarization modes. Thus, each cubic volume of $\left(\frac{2\pi}{L}\right) \times \left(\frac{2\pi}{L}\right) \times \left(\frac{2\pi}{L}\right)$ in the k-space corresponds to two radiation modes. For $L \gg \lambda$, the k-values can be considered continuous. The total number of radiation modes per physical volume L^3 from 0 to $|\vec{k}| \equiv k$ in the three-dimensional k-space in the spherical coordinate system is, therefore:

$$\frac{N}{L^3} \cong (2 \times \text{volume of sphere of radius } k) \bigg/ \left(\frac{8\pi^3}{L^3}\right) \times L^3$$

$$= \left(2 \times \frac{4\pi k^3}{3}\right) \bigg/ \left(\frac{8\pi^3}{L^3}\right) \times L^3 = \frac{k^3}{3\pi^2},$$

and the density-of-modes is:

$$D(\nu) = \frac{\partial}{\partial \nu}\left(\frac{N}{L^3}\right) = \frac{\partial}{\partial \nu}\left(\frac{\omega^3}{3\pi^2 c^3}\right) = \frac{8\pi \nu^2}{c^3}. \tag{5.53}$$

According to the theorem of equipartition of energy in classical statistical mechanics, the thermal energy $\langle E_{th} \rangle$ per mode of the radiation oscillator is $k_{\bar{B}} T$,

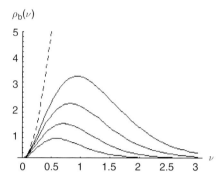

Figure 5.4. Black-body radiation spectrum. Dashed curve – Rayleigh–Jeans law ($T = 1600$ K ν in 10^{14} Hz and $\rho(\nu)$ in 10^{-16} ergHz^{-1}cm^{-3}). Solid curves – Planck's law. (Top to bottom: $T = 1600, 1400, 1200, 1000$ K.)

where $k_{\bar{B}}$ is the Boltzmann constant and is equal to 1.38×10^{-16} erg/K. Thus, the black-body radiation spectrum should, according to classical physics, be:

$$\rho_b(\nu) = D(\nu) \times \langle E_{th} \rangle = \frac{8\pi \nu^2}{c^3} k_B T, \tag{5.54}$$

which varies quadratically with the frequency ν and is proportional to the temperature of the radiation. It is known as the Rayleigh–Jeans law. It agrees very well with experiments in the low frequency range. It fails totally, however, in the high frequency limit; for it predicts that the thermal radiation energy increases as the frequency-squared indefinitely, as shown in Figure 5.4, which cannot possibly be correct physically. This anomaly is known as the "ultraviolet catastrophe." The correct explanation lies in the fact that the radiation must be quantized and $\langle E_{th} \rangle \neq k_B T$ equally for all the modes.

According to quantum mechanics, the average thermal energy per mode will depend on the frequency of the mode and the temperature, because the extent of thermal excitation in each mode is determined by the probability of occupation of the quantized energy levels of the radiation in that mode. Furthermore, the energy of each mode is not a continuous variable but is quantized corresponding to the eigen values $(n + \frac{1}{2})\hbar\omega$, (5.50a), of the Hamiltonian (5.50), where n is the number of photons due to thermal excitation in the mode and it is an integer equal to 0, 1, 2, 3, … The $\frac{1}{2}\hbar\omega$ term corresponds to the vacuum fluctuations associated with each linear polarization of the radiation mode. The probability of occupation due to thermal excitation of each of these quantized levels is:

$$P_n = \frac{e^{-nh\nu/k_B T}}{\sum\limits_{n=0,1,2,3\ldots} e^{-nh\nu/k_B T}} = e^{-nh\nu/k_B T}\left[1 - e^{-h\nu/k_B T}\right].$$

The average photon number per mode based on this probability distribution function is:

$$\langle n \rangle = \sum_n n\, P_n = \frac{1}{e^{h\nu/k_B T} - 1},$$

which is generally referred to as the "Bose–Einstein law." The corresponding average thermal energy per mode is then:

$$\langle E_{th} \rangle = \sum_n P_n n\, h\nu = \frac{h\nu e^{-h\nu/k_B T}}{1 - e^{-h\nu/k_B T}}.$$

Thus, the black-body spectrum according to quantum mechanics is:

$$\rho_b(\nu) = D(\nu) \times \langle E_{th} \rangle = \frac{8\pi h\nu^3}{c^3} \cdot \frac{1}{e^{h\nu/k_B T} - 1}, \tag{5.55}$$

which is also shown in Figure 5.4 and agrees precisely with the experimental results. There is no longer any "ultraviolet catastrophe," since $\lim_{\nu \to \infty} \rho_b(\nu) \to 0$. In the limit of low frequencies, (5.55) agrees exactly with the Rayleigh–Jeans law. Equation (5.55) is known as Planck's black-body radiation law, which was first derived empirically based on his postulate that the energy of light waves is quantized in units of $h\nu$ as "photons." The precise agreement of Planck's radiation law with observations was historically the first experimental proof of the validity of the concept of photons. This postulate has since been substantiated by numerous experiments including, for example, the all-important photoelectric effect.

Quantum theory of coherent optical states

From (5.48a) and (5.48b), the quantum mechanic operators representing the electric and magnetic fields of a monochromatic linearly polarized plane light wave are, respectively, of the forms:

$$\hat{\vec{E}}(\vec{r}, t) \equiv \hat{E}_x(z, t)\, \mathbf{e}_x = i\sqrt{\frac{2\pi\hbar\omega_k}{L}} \left(\hat{a}_k^- e^{-i\omega_k t + ikz} - \hat{a}_k^+ e^{i\omega_k t - ikz} \right) \mathbf{e_x}, \tag{5.56a}$$

and

$$\hat{\vec{B}}(\vec{r}, t) \equiv \hat{B}_y(z, t)\, \mathbf{e}_y = i\sqrt{\frac{2\pi\hbar\omega_k}{L}} \left(\hat{a}_k^- e^{-i\omega_k t + ikz} - \hat{a}_k^+ e^{i\omega_k t - ikz} \right) \mathbf{e_y}. \tag{5.56b}$$

Since $\hat{a}_k^+$ and $\hat{a}_k^-$ do not commute, there is no simultaneous eigen state of these two operators and one cannot specify simultaneously the complex amplitude and its complex conjugate of the E field or B field. One can, however, specify the intensity of the wave in terms of the photon number, or the eigen value n of the operator $\hat{a}_k^+ \hat{a}_k^-$

and the corresponding photon-number state $|n_k\rangle$. Many quantum optics problems can be studied using these states as the basis states. However, in such a state, the phase information is completely lost and it is difficult to compare the results expressed in the representation in which $\hat{a}_k^+ \hat{a}_k^-$ is diagonal directly with the results in the classical limit where the intensity and phase of the coherent optical wave are both known accurately, such as the output of an ideal single-mode laser.

A useful alternative approach is to use the eigen states $|\alpha_k\rangle$ of the operator representing the complex amplitude of the electromagnetic field, for example, $\hat{a}_k^-$ corresponding to the eigen values α_k:

$$\hat{a}_k^- |\alpha_k\rangle = \alpha_k |\alpha_k\rangle \tag{5.57}$$

as the basis states. From (5.56a) and (5.56b), the eigen value α_k is then proportional to the complex amplitude of the electric and magnetic fields. For reasons to be discussed in detail below, the $|\alpha_k\rangle$ state is known as the "coherent state," because it asymptotically approaches the state of a classical coherent electromagnetic wave with a well defined phase and amplitude as the average photon number increases. These $|\alpha_k\rangle$ states form a complete but not necessarily orthogonal set. A full quantum theory based on such a representation was developed by R. J. Glauber and first published in *Phys. Rev. Letters* **10**, 84 (1963) and *Phys. Rev.* **131**, 2766 (1963). Only a very brief introduction is given in what follows.

Measurement of the photon number when the electromagnetic wave is in a coherent state $|\alpha\rangle$ will not always yield the same result since $\hat{a}^-$ and $\hat{a}^+$ do not commute. In fact, one will find a statistical distribution of the photon numbers. This probability distribution can be found from the scalar product $|\langle n|\alpha\rangle|^2$, or the expansion coefficients of $|\alpha\rangle$ in terms of the basis states $|n\rangle$:

$$|\alpha\rangle = \sum_n \langle n|\alpha\rangle \, |n\rangle, \tag{5.58}$$

which can be readily obtained on the basis of the solutions of the harmonic oscillator problem already found in the previous sections. From (5.25) and (5.57), we know that:

$$\langle n|\alpha\rangle = \langle 0| \frac{(\hat{a}^-)^n}{\sqrt{n!}} |\alpha\rangle = \frac{(\alpha)^n}{\sqrt{n!}} \langle 0|\alpha\rangle. \tag{5.59}$$

Normalizing the state $|\alpha\rangle$ gives:

$$1 = |\langle\alpha|\alpha\rangle|^2 = \sum_n |\langle n|\alpha\rangle|^2$$

$$= |\langle 0|\alpha\rangle|^2 \sum_n \frac{\alpha^{2n}}{n!} = |\langle 0|\alpha\rangle|^2 e^{|\alpha|^2};$$

therefore,

$$\langle 0|\alpha\rangle = e^{-\frac{1}{2}|\alpha|^2}, \tag{5.60}$$

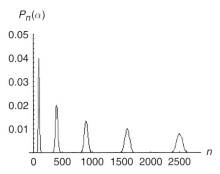

Figure 5.5.   Probability distribution functions (Poisson) of the photon numbers in the coherent optical states with: $|\alpha| = 10, 20, 30, 40$, and 50 (from left to right). Note that the area under each curve is equal to 1.

taking its inconsequential phase to be zero. Equation (5.60) shows that the ground state $|\alpha = 0\rangle$ in the coherent state representation is identical to the ground state $|n = 0\rangle$ in the photon-number state representation, which is the vacuum state; in both cases, there is no photon present beyond the vacuum fluctuations. If the magnitude of the complex amplitude is finite, the coherent state is in general a superposition of fixed-photon-number states, from (5.58) – (5.60):

$$|\alpha\rangle = e^{-\frac{1}{2}|\alpha|^2} \sum_n \frac{(\alpha)^n}{\sqrt{n!}} \, |n\rangle. \tag{5.61}$$

The photon number probability distribution for a beam in the $|\alpha\rangle$ state with an intensity:

$$I = |\alpha|^2 h\nu c/L = \bar{n} \, h\nu c/L, \tag{5.62}$$

is, from (5.61), the well-known "Poisson distribution":

$$P_n(\alpha) = |\langle n|\alpha\rangle|^2 = \frac{|\alpha|^{2n}}{n!} e^{-|\alpha|^2}, \tag{5.63}$$

with the mean photon number, $\bar{n} \equiv \langle \alpha | \hat{a}^+ \hat{a} | \alpha \rangle = |\alpha|^2$, per radiation mode in a volume of unit cross-sectional area and length L. Note that, from (5.62), for a long section $(L \gg \lambda)$ optical wave of duration $\tau = L/c$ and intensity I, the mean number of photons per mode per pulse of unit cross-sectional area is equal to:

$$\bar{n} = \frac{I\tau}{h\nu}, \tag{5.64}$$

as expected.

   The physical significance of the coherent state, (5.57) and (5.61), is that, in the limit of large $|\alpha|$, it represents a state of the fields approaching that of the classical coherent electromagnetic wave and α has the meaning of being the complex amplitude of the classical fields. This fact can, perhaps, best be appreciated qualitatively by looking at a few numerical examples based on the Poisson distribution function (5.63), as shown in Figure 5.5. Five numerical cases are shown corresponding to the $|\alpha|^2$ values of 10^2, 20^2, 30^2, 40^2, and 50^2. Although these mean photon numbers per mode are still very small, the trend is, however, clear from the numerical results. The average photon numbers $\bar{n}$ are indeed equal to the numerical values of $|\alpha|^2$. As the average photon number $\bar{n}$ increases, the numerical results show that, although the uncertainty increases, it increases far slower than the average photon number. Thus, the uncertainty relative to the average photon number decreases with increasing average photon numbers. Indeed, from (5.57), the average photon number is:

$$\bar{n} = |\langle\alpha|\hat{a}^+\hat{a}^-|\alpha\rangle|^2 = |\alpha|^2,$$

and, from (5.57) and (5.49), the absolute value of the uncertainty in the photon number is:

$$\Delta n = \sqrt{\langle\alpha|(\hat{a}^+\hat{a})^2|\alpha\rangle - (\langle\alpha|\hat{a}^+\hat{a}|\alpha\rangle)^2} = |\alpha| = (\bar{n})^{\frac{1}{2}},$$

which increases as $(\bar{n})^{1/2}$. Thus, from the uncertainty relationship between the phase and the photon number, the phase of the wave, $\Delta\phi$, decreases as $\sim(\bar{n})^{-1/2}$ and becomes better defined and approaches that of a classical coherent wave as the average photon number increases. On the other hand, the uncertainty in the photon number relative to the average value of the photon number $\Delta n/\bar{n}$ is also equal to $|\alpha|^{-1} = (\bar{n})^{-1/2}$ and decreases as $(\bar{n})^{-1/2}$. It means that, as the photon number increases, the intensity of the wave in the $|\alpha\rangle$ state becomes better and better defined as well, again as in a classical coherent wave.

In conclusion, for large photon numbers, the quantum coherent state approaches the limit of a classical coherent wave with a well defined phase. The corresponding intensity of the classical coherent wave is determined by the average photon number with a small relative uncertainty and the phase of the wave is well defined. In contrast, the fixed-photon-number state has a probability distribution function that is always a delta function in the photon number space. In other words, the intensity is specified with absolute certainty, but the phase is totally uncertain. To describe effects involving coherent waves with well defined phases quantum mechanically, one should, therefore, use the coherent state functions (5.61) as the basis states. This is a relatively new theory. It is important for understanding the statistical properties of the laser beam and for the new fields of atom lasers and quantum information science with possible applications in quantum computing and quantum cryptography.

5.5 Problems

5.1 Show that, for an eigen state of a one-dimensional harmonic oscillator, the following results are true:

(a) The expectation values of the position and momentum are zero.

(b) The expectation values of the potential energy and the kinetic energy are equal.

(c) The uncertainty product of the position and momentum $\Delta x \, \Delta p_x$ is equal to $(n + \frac{1}{2})\hbar$.

5.2 For a one-dimensional harmonic oscillator, give in the basis in which the Hamiltonian is diagonal the matrix representations of:

(a) the position and momentum operators $\hat{x}$ and $\hat{p}_x$, respectively;
(b) the operator products $\hat{a}^+ \hat{a}^-$ and $\hat{a}^- \hat{a}^+$.
(c) Using the matrices found in (b), show that the commutation relationship (5.12) is satisfied.

5.3 Show that the wave function of the form given in Eq. (5.32) indeed satisfies the time-independent Schrödinger equation for the one-dimensional harmonic oscillator.

5.4 Suppose the harmonic oscillator is initially in a superposition state $|\Psi(t=0)\rangle = \frac{1}{\sqrt{2}}[\, |0\rangle + |1\rangle]$, give the expectation value of the position of the oscillator $\langle x \rangle_t \equiv \langle \Psi(t)|x|\Psi(t)\rangle$ as a function of time.

5.5 Verify Eqs. (5.48a), (5.48b), (5.50), and (5.50a).

5.6 Give the Rayleigh–Jeans law and Planck's law for black-body radiation as functions of wavelength and in units of energy per volume per wavelength-interval, rather than in terms of frequency as in (5.54) and (5.55). Show explicitly that the corresponding Rayleigh–Jeans law shows that $\rho_b(\lambda) \, d\lambda$ is proportional to λ^{-4} in the wavelength-space and, therefore, diverges in the ultraviolet limit $\lambda \to 0$.

6 The hydrogen atom

The hydrogen atom is the Rosetta stone of the early twentieth century atomic physics. The attempt to decipher its structure and properties led to the development of quantum mechanics and the unraveling of many of the mysteries of atomic, molecular, and solid state physics, and a good deal of chemistry and modern biology. Unlike the various one-dimensional model problems that we have been studying in the previous chapters, the hydrogen atom is a real physical system in three dimensions. It consists of an electron moving in a spherically symmetric potential well due to the Coulomb attraction of the positively charged nucleus. In three dimensions, the electron is not constrained to move linearly. It can execute orbital motions and, thus, has angular momentum. Not only is the total energy of the electron in the atom quantized, its angular momentum also has interesting and unexpected quantized properties that cannot possibly be understood on the basis of classical mechanics and electrodynamics. They are, however, the natural and necessary consequences of the basic postulates of quantum mechanics, as will be shown in this chapter.

According to classical mechanics and electrodynamics, it is not possible to have a stable structure consisting of a small positively charged nucleus at the center of an electrically neutral atom with an electron sitting in its vicinity. For the electron not to be attracted into the positive charge, it must be orbiting around the nucleus so that the centrifugal force will counter the Coulomb attraction of the nucleus and maintain a constant electron orbit. Yet, if the electron is orbiting, it is being accelerated and must radiate and lose energy according to classical electrodynamics. Losing energy means it will slow down and eventually collapse into the nucleus. Thus, if quantum mechanics is to provide an explanation of how the electron and the nucleus can form a stable atom, it must show that the Coulomb potential well centered on the positively charged nucleus has stationary bound states with finite binding energies for the electron to occupy.

6.1 The Hamiltonian of the hydrogen atom

The model of the hydrogen atom being considered consists of an electron of negative charge $-e$ (4.803×10^{-10} esu) and mass m_e (0.91×10^{-27}g) and a nucleus with a positive charge of $+e$ and a much larger mass M equal to 1836 times m_e. Both are assumed to be point particles of infinitely small size. This two-particle (electron of mass m_e and nucleus of mass M) problem can be converted into a one-particle problem by considering the motion of the electron relative to that of the nucleus in the center-of-mass frame of the two particles according to the principles of classical mechanics. In this frame, the electron of mass m_e is replaced by a particle of "reduced mass" $\mu \equiv \dfrac{M m_e}{M + m_e}$ moving relatively to a nucleus at rest and fixed at the origin $(0,0,0)$ of, for example, a spherical coordinate system (r, θ, ϕ). To simplify the notation, in the following discussion of the hydrogen atom, we simply use 'm' ($\approx m_e$ for $M \gg m_e$) in place of the reduced mass μ.

The potential energy $V(r)$ of the electron due to the Coulomb attraction of the nucleus in free space is (unrationalized cgs units, $\varepsilon_0 = 1$):

$$V(r) = -\frac{e^2}{r}. \tag{6.1}$$

The corresponding Hamiltonian and the time independent Schrödinger equation are:

$$\hat{H}\Psi_E(r,\theta,\phi) = \left[-\frac{\hbar^2}{2m}\nabla^2 + V(r) \right] \Psi_E(r,\theta,\phi) = E\,\Psi_E(r,\theta,\phi), \tag{6.2}$$

which in the spherical coordinate system is:

$$\left\{ -\frac{\hbar^2}{2m}\left[\frac{1}{r^2}\frac{\partial}{\partial r}\left(r^2 \frac{\partial}{\partial r}\right) + \frac{1}{r^2 \sin^2\theta}\frac{\partial}{\partial\theta}\left(\sin\theta\frac{\partial}{\partial\theta}\right) + \frac{1}{r^2 \sin^2\theta}\frac{\partial^2}{\partial\phi^2} \right] - \frac{e^2}{r} \right\}\Psi_E(r,\theta,\phi)$$
$$= E\Psi_E(r,\theta,\phi). \tag{6.3}$$

The boundary conditions on the eigen functions are that $\Psi_E(r,\theta,\phi)$ must be finite and single valued at any spatial point ($0 \le r$, $0 \le \theta \le \pi$, $0 \le \phi \le 2\pi$). Since $|\Psi_E(r,\theta,\phi)|^2$ is the probability distribution function corresponding to a bound state, it must be square integrable to unity, i.e. normalizable: $\iiint |\Psi_E(r,\theta,\phi)|^2 r^2 dr \sin\theta\, d\theta\, d\phi = 1$.

At this point, one can proceed to solve the time-independent Schrödinger equation (6.3) by the standard method of separation of variables and find the eigen functions and eigen values. It will, however, involve a great deal of mathematical details without offering much insight. We will postpone doing so until Section 6.3. Instead, we will try to reach some conclusions about the angular momentum properties first. The results will greatly facilitate the solution of (6.3) later.

6.2 Angular momentum of the hydrogen atom

The theory of angular momentum plays a crucial role in the understanding of the structure and properties of atoms, molecules, and solids. The hydrogen atom is a simple model with which to introduce some of the elementary concepts of the theory.

The classical expression of the orbital angular momentum of a point particle is: $\vec{L} = \vec{r} \times \vec{p}$. Therefore, the corresponding quantum mechanical operator in the Schrödinger representation is:

$$\hat{\vec{L}} = \hat{\vec{r}} \times \hat{\vec{p}} = \hat{\vec{r}} \times (-i\hbar\nabla), \tag{6.4}$$

or

$$\hat{L}_x = -i\hbar\left(y\frac{\partial}{\partial z} - z\frac{\partial}{\partial y}\right), \tag{6.4a}$$

$$\hat{L}_y = -i\hbar\left(z\frac{\partial}{\partial x} - x\frac{\partial}{\partial z}\right), \tag{6.4b}$$

$$\hat{L}_z = -i\hbar\left(x\frac{\partial}{\partial y} - y\frac{\partial}{\partial x}\right). \tag{6.4c}$$

A total orbital angular momentum operator can also be defined and it is:

$$\hat{L}^2 \equiv \hat{L}_x^2 + \hat{L}_y^2 + \hat{L}_z^2. \tag{6.5}$$

It follows from (2.11a & b) and (6.4a, b, c) that the components of the orbital angular momentum operator satisfy the cyclic commutation relationships:

$$[\hat{L}_x, \hat{L}_y] = i\hbar\,\hat{L}_z, \quad [\hat{L}_y, \hat{L}_z] = i\hbar\,\hat{L}_x, \quad [\hat{L}_z, \hat{L}_x] = i\hbar\,\hat{L}_y; \tag{6.6}$$

and

$$[\hat{L}^2, \hat{L}_x] = 0, \quad [\hat{L}^2, \hat{L}_y] = 0, \quad [\hat{L}^2, \hat{L}_z] = 0. \tag{6.7}$$

Note that, instead of the x and y components of the orbital angular momentum, we can also define a right and a left circular component in the (xy) plane of the angular momentum as:

$$\hat{L}_+ = \hat{L}_x + i\hat{L}_y \quad \text{and} \quad \hat{L}_+ = \hat{L}_x - i\hat{L}_y. \tag{6.8}$$

The corresponding commutation relations are:

$$[\hat{L}_z, \hat{L}_+] = \hbar\,\hat{L}_+, \quad [\hat{L}_z, \hat{L}_-] = -\hbar\,\hat{L}_-, \quad [\hat{L}_+, \hat{L}_-] = 2\hbar\,\hat{L}_z; \tag{6.9}$$

and

$$[\hat{L}^2, \hat{L}_\pm] = 0. \tag{6.10}$$

These commutation relations have the very important implication that the total orbital angular momentum of the electron can be specified simultaneously with one and only one of the three components of the orbital angular momentum, because of (6.6), (6.7), and (6.9), but the choice of which one is arbitrary. The chosen component is the one in the direction of an arbitrarily chosen "axis of quantization," for reasons that will become clear later. By convention, the z axis is usually chosen arbitrarily as the axis of quantization. With this choice, it means physically that, in general, the magnitude of a finite orbital angular momentum vector of the electron and its projection along the axis of quantization can be precisely specified, but its particular direction in the xy plane can not be specified, because the x and y components of the vector are totally uncertain. Thus, the orbital angular momentum vector must lie on the surface of a cone with the axis of quantization as the symmetry axis and its apex at (0, 0, 0). The cosine of the half-apex angle is equal to the ratio of the projection of the vector along the symmetry axis to the magnitude of the angular momentum vector. A more detailed discussion of this point will be given later when we show that both the magnitude of this vector and its projection along the axis of quantization are quantized. It should be pointed out, however, if the particle is in the particular state where it does not possess any angular moment, then all three components can be specified as precisely zero with certainty (see, for example, the discussion immediately following (2.12)).

Knowing that the total orbital angular momentum and its projection along the axis of quantization can be specified precisely at the same time means that the electron can be in a state that is a simultaneous eigen state of $\hat{L}^2$ and $\hat{L}_z$. Let us first find the eigen states and eigen values of $\hat{L}_z$, which in the Schrödinger representation in the spherical coordinate system is, from (6.4c), simply:

$$\hat{L}_z = -i\hbar \frac{\partial}{\partial \phi}. \tag{6.11}$$

The corresponding eigen value equation is:

$$\hat{L}_z \Phi_{\ell_z}(\phi) = \ell_z \Phi_{\ell_z}(\phi), \tag{6.12a}$$

or

$$-i\hbar \frac{\partial}{\partial \phi} \Phi_{\ell_z}(\phi) = \ell_z \Phi_{\ell_z}(\phi), \tag{6.12b}$$

Since the general boundary condition on the wave functions is that they must be finite and single-valued at any spatial point, the value of the wave function at any value of ϕ and $(\phi + 2N\pi)$, where N is an integer $= \pm 0, \pm 1, \pm 2, \pm 3, \ldots$, must be the same, or more explicitly:

$$\Phi_{\ell_z}(\phi) = \Phi_{\ell_z}(\phi + 2N\pi). \tag{6.13}$$

To satisfy (6.12b), $\Phi_{\ell_z}(\phi)$ must be of the form:

$$\Phi_{\ell_z}(\phi) = Ce^{\frac{i\ell_z}{\hbar}\phi}. \tag{6.14}$$

To satisfy the boundary condition (6.13), the eigen value must be:

$$\ell_z = m_\ell \hbar, \quad \text{where } m_\ell = 0, \pm 1, \pm 2, \pm 3, \ldots, \tag{6.15}$$

therefore, the corresponding normalized eigen function of $\hat{L}_z$, or the "azimuthal angular momentum," must be of the form:

$$\Phi_{m_\ell}(\phi) = \frac{1}{\sqrt{2\pi}} e^{im_\ell \phi}. \tag{6.16}$$

These results have profound physical implications: Eqs. (6.12a) and (6.15) show that the z component of the orbital angular momentum must be quantized and in units of $\pm \hbar$. This is a concept that is totally absent in classical mechanics. Historically, the conjecture by Bohr and Sommerfeld that the angular momentum of the atom might have to be quantized was one of the first hints from Nature that a totally new kind of physics might be needed to understand the structure of atoms. From the point of view of quantum mechanics, the reason the azimuthal angular momentum must be quantized is that the corresponding eigen state of the electron is a de Broglie wave circulating in the ϕ direction around the axis of quantization, as shown in Eqs. (6.14). Because the wave function must be single-valued in space, it must satisfy a periodic boundary condition on ϕ as in (6.13). Much like the reason why the linear momentum of a particle in a box must be quantized because of the boundary conditions at the walls defining the spatial region to which the particle is confined, the boundary condition relating the value of $\Phi(\phi)$ at $\phi = 0$ and 2π leads to the quantization of the angular momentum. The reason it is quantized in units of $\pm \hbar$ is related to the basic commutation relationships, (2.11 a & b), and can be viewed as a consequence of Heisenberg's uncertainty principle. Note that there is also a very subtle point involving the analogy between the quantization of the angular momentum and the linear moment of a particle in a box. In the case of a particle in a box of impenetrable walls, the quantized linear momentum has an uncertainty associated with it because the wave function is defined only within the box of finite width. For the angular momentum, the boundary condition (6.13) is a periodic one with no restriction on the value of ϕ, which can be from $-\infty$ to $+\infty$, not just from 0 to 2π. It means that the uncertainty in ϕ is unlimited and, hence, the corresponding azimuthal angular momentum can have sharply defined quantized values of $m\hbar$. All this may sound a little bizarre from the view point of classical mechanics. Yet, as we will see later, all these predictions based on quantum mechanics agree perfectly well with numerous results of the most sophisticated experiments, while classical mechanics would have missed all of it.

We can now try to find the simultaneous eigen state of the operators $\hat{L}^2$ and $\hat{L}_z$, which must be a function of both θ and ϕ. Let us designate such an eigen function by $Y_L(\theta, \phi)$. It must satisfy the eigen value equation:

$$\hat{L}^2 Y_L(\theta, \phi) = L^2 Y_L(\theta, \phi) \tag{6.17}$$

with the corresponding eigen value L^2, which is a number yet to be determined. In the spherical coordinate system, (6.17) becomes:

$$\left[-\hbar^2 \left(\frac{1}{\sin\theta} \frac{\partial}{\partial\theta} \sin\theta \frac{\partial}{\partial\theta} + \frac{1}{\sin^2\theta} \frac{\partial^2}{\partial\phi^2} \right) \right] Y_L(\theta, \phi) = L^2 Y_L(\theta, \phi). \tag{6.18}$$

Since $Y_L(\theta, \phi)$ is a simultaneous eigen function of $\hat{L}_z$ and $\hat{L}^2$, it must be proportional to the eigen function of $\hat{L}_z$, or $\Phi_{m_\ell}(\phi)$, and the proportionality factor must be independent of ϕ but a function of θ only. Thus, $Y_L(\theta, \phi)$ must depend on m_ℓ and must be separable into products of two factors, one involving θ only, and the other involving ϕ only:

$$Y_{L m_\ell}(\theta, \phi) = \Theta_{L m_\ell}(\theta) \Phi_{m_\ell}(\phi). \tag{6.19}$$

Substituting (6.16) and (6.19) into (6.18) gives the eigen value equation for $\Theta_{L m_\ell}(\theta)$:

$$\left[\frac{1}{\sin\theta} \frac{\partial}{\partial\theta} \sin\theta \frac{\partial}{\partial\theta} - \frac{m_\ell^2}{\sin^2\theta} \right] Y_{L m_\ell}(\theta, \phi) = -\frac{L^2}{\hbar^2} Y_{L m_\ell}(\theta, \phi). \tag{6.20}$$

Here again, just like in the case of Schrödinger's equation for the harmonic oscillator considered in Chapter 5, we have a differential equation with complicated variable coefficients. Fortunately, this differential equation is related to the *Legendre equation* and its solutions are well known as the "*spherical harmonics*." There is no need for us to "reinvent the wheel" here. We will just quote the known results, and discuss their physical significance. It is well known that Eq. (6.20) will only have solutions that satisfy the boundary condition that the wave functions are finite and single-valued everywhere spatially, if the eigen values are of the form:

$$L^2 = \ell(\ell + 1)\hbar^2, \tag{6.21}$$

where the orbital angular momentum is quantized and the orbital quantum number ℓ is equal to:

$$\ell = 0, 1, 2, 3, \ldots, \tag{6.22}$$

and the magnitude of the azimuthal quantum number m_ℓ in (6.15) and (6.16) is limited to less than or equal to ℓ:

$$|m_\ell| = 0, \ 1, \ 2, \ 3, \ \ldots \leq \ell. \tag{6.23}$$

The corresponding eigen functions are proportional to the well-known associated Legendre functions of the form $P_\ell^{m_\ell}(\cos\theta)$, which are all tabulated. The complete normalized eigen functions $Y_{\ell m_\ell}(\theta, \phi)$, with the label in the subscript now changed from L to ℓ, of $\hat{L}^2$ are the spherical harmonics:

$$Y_{\ell m_\ell}(\theta, \phi) = \sqrt{\frac{(2\ell + 1)(\ell - m_\ell)!}{4\pi(\ell + m_\ell)!}} (-1)^{m_\ell} P_\ell^{m_\ell}(\cos\theta) e^{im_\ell\phi}; \tag{6.24}$$

thus,

$$\hat{L}^2 Y_{\ell m_\ell}(\theta, \phi) = \ell(\ell + 1)\hbar^2 Y_{\ell m_\ell}(\theta, \phi), \tag{6.25}$$

$$\hat{L}_z Y_{\ell m_\ell}(\theta, \phi) = m_\ell \hbar Y_{\ell m_\ell}(\theta, \phi), \tag{6.26}$$

and

$$\int\limits_0^{2\pi} \int\limits_0^\pi Y_{\ell m_\ell}(\theta, \phi) Y_{\ell' m_\ell'}(\theta, \phi) \sin\theta \, d\theta \, d\phi = \delta_{\ell\ell'} \delta_{m_\ell m_\ell'}. \tag{6.27}$$

The associated Legendre function can be generated from the Legendre polynomial $P_\ell^{m_\ell}(\xi)$ defined as follows:

$$P_\ell^{m_\ell}(\xi) = \frac{1}{2^\ell \ell!} (1 - \xi^2)^{\frac{m_\ell}{2}} \frac{\partial^{\ell+m_\ell}}{\partial \xi^{\ell+m_\ell}} (\xi^2 - 1)^\ell, \tag{6.28}$$

where $\xi = \cos\theta$ and ℓ is a positive integer $0, 1, 2, \ldots$ The associated Legendre function and the spherical harmonics are all well-known and tabulated. The first few of the associated Legendre functions are:

$$P_0^0(\xi) = 1, \quad P_1^0(\xi) = \xi, \quad P_1^1(\xi) = \sqrt{1 - \xi^2},$$
$$P_2^0(\xi) = \frac{1}{2}(3\xi^2 - 1), \quad P_2^1(\xi) = 3\xi\sqrt{1 - \xi^2}, \quad P_2^2(\xi) = 3(1 - \xi^2). \tag{6.28a}$$

The first few spherical harmonics are listed in Table 6.1.

Table 6.1. *Examples of spherical harmonics* $Y_{\ell m_\ell}(\theta, \phi)$

$$Y_{00} = \frac{1}{\sqrt{4\pi}}, \quad Y_{10} = \sqrt{\frac{3}{4\pi}} \cos\theta = \sqrt{\frac{3}{4\pi}} \frac{z}{r},$$

$$Y_{1\pm1} = \mp\sqrt{\frac{3}{8\pi}} e^{\pm i\phi} \sin\theta = \mp\sqrt{\frac{3}{8\pi}} \frac{x \pm iy}{r},$$

$$Y_{20} = \sqrt{\frac{5}{16\pi}}(3\cos^2\theta - 1) = \sqrt{\frac{5}{16\pi}} \frac{2z^2 - x^2 - y^2}{r^2},$$

$$Y_{2\pm1} = \mp\sqrt{\frac{15}{8\pi}} e^{\pm i\phi} \cos\theta \sin\theta = \mp\sqrt{\frac{15}{8\pi}} \frac{(x \pm iy)z}{r^2},$$

$$Y_{2\pm2} = \sqrt{\frac{15}{32\pi}} e^{\pm i2\phi} \sin^2\theta = \sqrt{\frac{15}{32\pi}} \frac{(x \pm iy)^2}{r^2}.$$

A word about the parity of the spherical harmonics. By inverting the coordinate axes through the origin $\vec{r} \to -\vec{r}$, or through the transformation $\phi \to \phi + \pi$ and $\theta \to \pi - \theta$, it can be shown on the basis of (6.16), (6.24), and (6.28) that the spherical harmonics transform as $Y_{\ell m_\ell}(\theta, \phi) \to (-1)^\ell Y_{\ell m_\ell}(\theta, \phi)$. Therefore, the parity of the spherical harmonics is even or odd according to whether ℓ is even or odd. The parity of the stationary states of the atom will have interesting consequences in the consideration of the interaction of atoms with electromagnetic fields, as will be discussed later. It is, therefore, of fundamental importance to such applications as the optical absorption or emission process.

In working with the eigen value equations and the functions of $\hat{L}^2$ and $\hat{L}_z$, the more efficient and compact Dirac's notation is often used. Thus, Eqs. (6.25)–(6.27) can also be written as:

$$\hat{L}^2|\ell m_\ell\rangle = \ell(\ell+1)\hbar^2|\ell m_\ell\rangle, \tag{6.25a}$$

$$\hat{L}_z|\ell m_\ell\rangle = m_\ell\hbar|\ell m_\ell\rangle, \tag{6.26a}$$

and

$$\langle \ell m_\ell | \ell' m'_\ell \rangle = \delta_{\ell\ell'}\delta_{m_\ell m'_\ell}. \tag{6.27a}$$

As is obvious, much of the information that is superfluous and repeated, such as the symbols Y, θ, ϕ, and the complicated integral, is not shown in Dirac's notation. In the same notation, the matrix representations in which $\hat{L}^2$ and $\hat{L}_z$ are diagonal are simply:

$$\langle \ell m_\ell | \hat{L}^2 | \ell' m'_\ell \rangle = \ell(\ell+1)\hbar^2\delta_{\ell\ell'}\delta_{m_\ell m'_\ell}, \tag{6.29}$$

and

$$\langle \ell m_\ell | \hat{L}_z | \ell' m'_\ell \rangle = m_\ell\hbar\delta_{\ell\ell'}\delta_{m_\ell m'_\ell}. \tag{6.30}$$

It can be shown from the properties of the spherical harmonics that the corresponding matrix representations of the circular components of the angular momentum operator, $\hat{L}_+$ and $\hat{L}_-$ defined in (6.8), are:

$$\langle \ell m_\ell | \hat{L}_\pm | \ell' m'_\ell \rangle = [(\ell \mp m_\ell)(\ell \pm m_\ell + 1)]^{\frac{1}{2}}\hbar\delta_{\ell\ell'}\delta_{m_\ell,(m'_\ell\pm 1)}. \tag{6.31}$$

Note that $\hat{L}_+$ and $\hat{L}_-$ have the same general form as the creation and annihilation operators $\hat{a}_+$ and $\hat{a}_-$ defined in connection with the harmonic oscillator problem studied in Chapter 5. Indeed, they have the same physical implications: $\hat{L}_+$ and $\hat{L}_-$ applied to the state $|\ell m_\ell\rangle$ change it into a state with one more or one fewer $\hbar$ of azimuthal angular momentum, respectively, in its projection along the axis of quantization.

These results on the eigen states of $\hat{L}^2$ and $\hat{L}_z$ give specific information on the orbital motion of the electron in the hydrogen atom. The length of the orbital angular

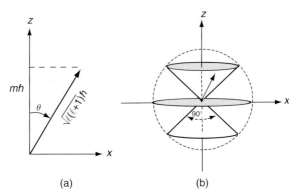

(a)   (b)

Figure 6.1. (a) An orbital angular momentum vector $\vec{L}$ of length $\sqrt{\ell(\ell+1)}\hbar$ with its projection $m\hbar$ on the axis of quantization, the z-axis. (b) Orbital angular momentum vectors $\vec{L}$ corresponding to the $\ell = 1$ and $m_\ell = \pm 1$ and 0 states are *uniformly distributed over the surfaces of cones* centered on the z-axis with an apex angle of 90^0 and the circular plane-surface in the (xy)-plane, respectively.

momentum vector of the electron and its projection on the axis of quantization must be quantized according to Eqs. (6.25) and (6.26). The vectors themselves must lie on the surfaces of cones centered on the axis of quantization with specific apex angles $2\theta_{\ell m_\ell}$ depending on the quantum numbers ℓ and m_ℓ: $\theta_{\ell m_\ell} = \cos^{-1}\frac{m_\ell}{\sqrt{\ell(\ell+1)}}$, as shown in Figure 6.1. For each set of ℓ and m_ℓ values, the projection of the angular momentum vector $\vec{L}$ in the xy plane is always independent of ϕ and, therefore, the vector has equal probability of being in any particular direction.

6.3 Solution of the time-independent Schrödinger equation for the hydrogen atom

With the eigen value equations for $\hat{L}^2$ and $\hat{L}_z$ solved and the corresponding eigen functions and eigen values at hand, the solution of the time-independent Schrödinger equation for the hydrogen atom is greatly simplified. By making use of Eq. (6.20), the corresponding Hamiltonian, (6.3), can also be written as:

$$\hat{H} = -\frac{\hbar^2}{2m}\left[\frac{1}{r^2}\frac{\partial}{\partial r}\left(r^2\frac{\partial}{\partial r}\right) - \frac{\hat{L}^2}{\hbar^2 r^2}\right] - \frac{e^2}{r}. \qquad (6.32)$$

Thus,

$$[\hat{H}, \hat{L}^2] = 0, \qquad [\hat{H}, \hat{L}_z] = 0, \qquad (6.33)$$

and the simultaneous eigen functions of $\hat{L}^2$ and $\hat{L}_z$ must be simultaneously the eigen functions of the Hamiltonian, $\hat{H}$, also. This means that all the results we have obtained in the previous section (6.2) on the orbital angular momentum apply also to the stationary states of the hydrogen atom, and the magnitude and one component of the orbital angular momentum are constants of motion. The solutions, $\Psi_E(r, \theta, \phi)$, of the Schrödinger equation:

$$\left\{ -\frac{\hbar^2}{2m} \left[\frac{1}{r^2} \frac{\partial}{\partial r} \left(r^2 \frac{\partial}{\partial r} \right) - \frac{\hat{L}^2}{\hbar^2 r^2} \right] - \frac{e^2}{r} \right\} \Psi_E(r, \theta, \phi) = E \Psi_E(r, \theta, \phi) \qquad (6.34)$$

must then be proportional to $Y_{\ell m_\ell}(\theta, \phi)$ and of the form:

$$\Psi_E(r, \theta, \phi) = R_E(r) Y_{\ell m_\ell}(\theta, \phi). \qquad (6.35)$$

Substituting (6.35) into (6.34) shows that the "radial wave function" $R_E(r)$ and E depend on the quantum number ℓ and satisfy the equation:

$$\left\{ -\frac{\hbar^2}{2m} \left[\frac{1}{r^2} \frac{\partial}{\partial r} \left(r^2 \frac{\partial}{\partial r} \right) \frac{\partial}{\partial r} - \frac{\ell(\ell+1)}{r^2} \right] - \frac{e^2}{r} \right\} R_{E_\ell}(r) = E_\ell R_{E_\ell}(r). \qquad (6.36)$$

Again, this is an ordinary differential equation with variable coefficients and can be solved by the standard method of power series expansion. It has been solved by mathematicians years ago and the solutions are now well known. There is no point in repeating the steps here. We will simply quote the results and concentrate on its physical implications.

First, the eigen value E_ℓ is independent of the azimuthal quantum number m_ℓ. Therefore, the energy level E_ℓ is degenerate with "$(2\ell + 1)$-fold orbital degeneracy." Physically, this spatial degeneracy is due to the fact that, because the probability distribution functions of the electron in all these $|\ell m_\ell\rangle$ states are independent of the azimuthal angle ϕ, the *potential* energy of the electron in the spherical Coulomb potential of the nucleus can not be different for these states. The *kinetic* energy of these $(2\ell + 1)$ states are also the same because they all have the same orbital quantum number ℓ. It is, therefore, expected that these states would have the same *total* energy and must be degenerate. Such a $(2\ell + 1)$-fold degeneracy is call "normal degeneracy"; it is a consequence of the spherical symmetry of the atom. For the particular case of Coulomb potential of the form $-e^2/r$, but not for spherical potential of any other form, it so happens that mathematically the eigen value E_ℓ is also independent of orbital quantum number ℓ. Thus, for the hydrogenic model that includes only the Coulomb interaction between the electron and the nucleus, the energy levels are also degenerate with respect to the orbital angular momentum quantum number ℓ. This degeneracy is called "accidental degeneracy." This accidental degeneracy of the hydrogen atom will be removed when other smaller effects that are neglected in the simple model used here are taken into account.

In an atom, the electron is bound to the nucleus. Of particular interest here are the "bound states" of the atom in which the electron energy is negative relative to that of a free electron, so that it is confined within the Coulomb potential well of the electron in the presence of the positively charged nucleus. These are analogous to the bound states of a finite square well potential case considered in Section 4.4. Solving Eq. (6.36) by the standard method of power series expansion shows that the Hamiltonian of the hydrogen atom has negative quantized eigen values:

$$E_n = -\frac{\hbar^2}{2ma_0^2 n^2} = \frac{E_1}{n^2}, \tag{6.37}$$

where $a_0 \equiv \dfrac{\hbar^2}{me^2}$ is the "Bohr radius" and is equal to 0.529 Å. E_1 is the ground-state energy of the hydrogen atom. Furthermore, the "principal quantum number" $n = 1, 2, 3, \dots, \infty$, and the orbital quantum number ℓ now must be less than or equal to $(n-1)$, or $\ell = 0, 1, 2, 3, \dots (n-1)$, so that the corresponding eigen functions $\Psi_{n\ell m}(r, \theta, \phi)$ are finite for all values of $r \geq 0$ and square integrable from $r = 0$ to ∞. These are the bound states of the hydrogen atom. Note the interesting similarities, and the differences, of the form of the quantized energies of the Coulomb potential well case, (6.37), with that of the infinite square well potential case given in (4.26). Here the n^2 term appears in the denominator of E_n rather than in the numerator as in (4.26), and the Bohr radius plays the role of d/π in the square well potential case. Thus, the quantized energy levels for the hydrogen atom come closer and closer together rapidly as the energy increases and merge into a continuum at $E_\infty = 0$ and above, as we anticipated previously on the basis of the quantized energy level structure of the simple square-well potential model discussed in Section 4.4.

Each of the bound states is specified by three "quantum numbers": $|n, \ell, m_\ell\rangle$. Thus, each energy level corresponding to the principal quantum number n has a total of

$$\sum_{\ell=0}^{n-1} (2\ell + 1) = n^2$$

fold orbital degeneracy. The zero energy level $E_n = 0$, for the state $n = \infty$, is the "ionization limit" of the atom. For $E > 0$, the electron has a positive kinetic energy relative to the ionization limit throughout three-dimensional space in the presence of a positively charged hydrogen ion.

Numerically, the ground-state $(n = 1)$ energy E_1 of the hydrogen atom is -13.6 eV $(1\text{eV} = 1.6019 \times 10^{-12}\text{erg})$ below the ionization limit $E_\infty = 0$. In other words, it takes a minimum of this amount of energy to free the electron from a hydrogen atom in its ground state. Since this minimum eigen value of the Hamiltonian is negative but finite, the electron can remain in this stable stationary ground state forever and never collapse into the nucleus. According to classical mechanics, this is not possible. Quantum mechanically, it is.

The r-dependent part of the eigen function or the normalized radial wave function $R_{n\ell}(r)$ is related to the so-called "associated Laguerre functions," $L_{n+\ell}^{2\ell+1}(\zeta)$:

$$R_{n\ell}(r) = A_{n\ell}e^{-\zeta/2}\zeta^\ell L_{n+\ell}^{2\ell+1}(\zeta),$$

where $\zeta \equiv \dfrac{2}{na_0}r$, and $A_{n\ell}$ is a normalization constant. The details of the derivation and the exact general forms of $L_{n+\ell}^{2\ell+1}(\zeta)$ and this normalization constant are not important

Table 6.2. *Examples of normalized radial wave functions $R_{n\ell}(r)$ of the hydrogen atom*

$$R_{10}(r) = a_0^{-3/2} 2 e^{-r/a_0}$$

$$R_{20}(r) = (2a_0)^{-3/2} 2 \left(1 - \frac{r}{2a_0}\right) e^{-r/2a_0}$$

$$R_{21}(r) = (2a_0)^{-3/2} \frac{1}{\sqrt{3}} \frac{r}{a_0} e^{-r/2a_0}$$

$$R_{30}(r) = (3a_0)^{-3/2} 2 \left[1 - \frac{2}{3}\frac{r}{a_0} + \frac{2}{27}\left(\frac{r}{a_0}\right)^2\right] e^{-r/3a_0}$$

$$R_{31}(r) = (3a_0)^{-3/2} \frac{4\sqrt{2}}{3} \frac{r}{a_0} \left(1 - \frac{r}{6a_0}\right) e^{-r/3a_0}$$

$$R_{32}(r) = (3a_0)^{-3/2} \frac{2\sqrt{2}}{27\sqrt{5}} \left(\frac{r}{a_0}\right)^2 e^{-r/3a_0}$$

except in detailed numerical work. When needed, they can always be found in the literature. For the present discussion, it is more informative to see a few explicit examples and consider their physical implications. The first few $R_{n\ell}(r)$ are listed in Table 6.2.

6.4 Structure of the hydrogen atom

Knowing the energy eigen values and eigen functions of the hydrogen atom, we can now describe and show schematically some of the structural properties of the atom in its ground and excited states.

First, the energy levels of the hydrogen atom relative to the Coulomb potential are shown qualitatively in Figure 6.2, and the various orbital degenerate states for each energy level are shown in Figure 6.3.

As a matter of notation, the orbital angular momentum states with $\ell = 0, 1, 2, 3, 4, 5, 6, \ldots$ were often referred to by the spectroscopists as the s (for "sharp"), p (for "principal"), d (for "diffused"), f (for "fundamental"), g, h, i, $\ldots$ states, respectively, before quantum mechanics was fully developed; this is now a widely used convention. For example, the $|n = 1, \ell = 0\rangle$ and $|n = 3, \ell = 2\rangle$ levels are commonly referred to as the 1s and 3d levels, respectively.

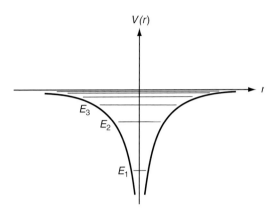

Figure 6.2. Schematic of the Quantized Energy Levels of a Coulomb Potential Well.

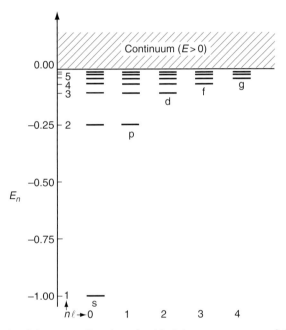

Figure 6.3. Schematic of the energy Levels and orbital degenerate states of the hydrogen atom (E_n in units of 13.6 eV).

Note that the 2s and 2p levels of the hydrogen atom are degenerate accidentally because of the particular form of the Coulomb potential in the simple model used for the atom. For other forms of spherically symmetric potentials, the energy eigen values will depend on the orbital angular momentum quantum number ℓ and this accidental degeneracy will be lifted. In addition, even in the case of Coulomb potential, if such subtle and small effects as, for example, the relativistic corrections and other corrections due to the interaction of the electron with its own field ("Lamb shift") (see, for example, Bethe and Jackiw (1986)) are taken into account, there is a small shift of the 2p from the 2s level. Such considerations go beyond the scope of this book.

Due to the spherical symmetry of the Coulomb potential, the states with different m_ℓ values but the same n and ℓ values are normally degenerate. This model neglects, however, the effects of the spinning motion of the electron. If the additional magnetic interaction between the spinning and the orbital motions of the electron is taken into account, this normal degeneracy also will be lifted, partially at least. Thus, for example, the 2p level of hydrogen will further split into two groups of states with different total spin–orbit coupled angular momentum values, as will be shown in Section 6.5.

The probability distribution of the electron in various stationary states depends on the wave function $\Psi_{n\ell m_\ell}(r, \theta, \phi) = R_{n\ell}(r) Y_{\ell m_\ell}(\theta, \phi)$ of the hydrogen atom. In the lowest energy state, or the ground-state $\Psi_{100}(r, \theta, \phi)$, $Y_{00}(\theta, \phi)$ is spherically symmetric and independent of θ and ϕ. The probability of finding the electron in the spherical shell between r and $r + dr$ is:

$$\int_0^{2\pi} \int_0^\pi |R_{10}(r)|^2 r^2 dr \sin\theta \, d\theta \, d\phi = 4\pi |R_{10}(r)|^2 r^2 dr,$$

which is shown schematically in Figure 6.4. It can be shown that the peak of this distribution function is exactly at the Bohr radius $r = a_0$. The electron in the stationary ground state can, therefore, be qualitatively visualized as being angularly uniformly distributed in some sort of spherical shell of radius a_0 around the nucleus in the hydrogen atom. Also, quantum mechanically, in the ground state $\ell = 0$ and there is no orbital angular momentum.

The probability distribution functions $|R_{n\ell}(r)|^2 r^2$ corresponding to some of the radial wave functions of the hydrogenic states $|n, \ell, m_\ell\rangle$ tabulated in Table 6.1 are shown in Figure 6.4. Note that the major peaks and average values of the distribution functions of the states with the same principal quantum number tend to cluster around each other and fall within the same general shell region radially. Within each such a shell, however, the distribution functions for states with different angular momentum quantum numbers ℓ and m_ℓ are very different. A few examples of the angular dependence of some of these are shown schematically in Figure 6.5.

In the chemistry literature, the wave functions are often referred to as the "orbitals," since they describe the orbital motions of the electrons in the atoms. Thus, the distribution functions shown in Figure 6.5 correspond to the s orbital and the p_z and the $p_{\pm 1}$ orbitals. Note that the donut-shaped probability distribution functions corresponding to the $p_{\pm 1}$ orbitals are independent of the azimuthal angle ϕ. We can also form p_x and p_y orbitals from the linear combinations of the $p_{\pm 1}$ orbitals:

$$|p_x\rangle = -\frac{1}{\sqrt{2}}(|p_{+1}\rangle - |p_{-1}\rangle) \quad \text{and} \quad |p_y\rangle = i(|p_{+1}\rangle + |p_{-1}\rangle) \tag{6.38}$$

In this case, the probability distribution functions for the p_x and p_y orbitals have exactly the same shape as that for the p_z orbital except that they are now pointed in the x and y directions, respectively, rather than in the z direction as in the former case. The shapes and the orientations of the atomic orbitals are of fundamental importance in

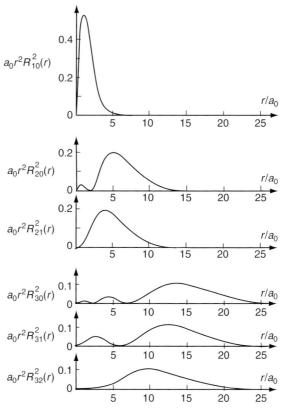

Figure 6.4. Examples of the radial probability distribution functions of the 1s, 2s, 2p, 3s, 3p and 3d states of the hydrogen atom.

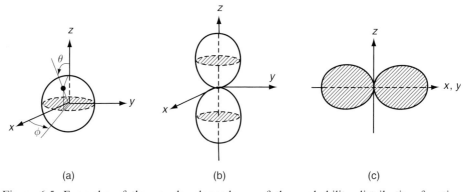

(a)   (b)   (c)

Figure 6.5. Examples of the angular dependence of the probability distribution functions $|Y_{\ell m_\ell}(\theta, \phi)|^2$: (a) $\ell = 0$, $m_\ell = 0$; (b) $\ell = 1$, $m_\ell = 0$; (c) $\ell = 1$, $m_\ell = \pm 1$, in this case the cross sections of the donut-shaped $|Y_{1,\pm 1}(\theta, \phi)|^2$ in the xz– and yz– planes are shown.

determining the basic structures and properties of molecules and solids formed from the constituent atoms, as we shall see later.

One can also gain some degree of qualitative understanding of the properties of the orbital angular momentum of the atom on the basis of the geometry of the wave functions. The probability distribution function corresponding to the s-state is spherically symmetric. An electron in such a state cannot be moving in any orbital trajectory, or posses any orbital angular momentum. Thus, the orbital angular momentum quantum number ℓ and the magnitude of the corresponding angular momentum vector $\sqrt{\ell(\ell+1)}\hbar$ are expected to be zero. The kinetic energy of the electron in the ground state is, therefore, due to the radial motion of the electron trapped in the Coulomb potential well.

For the p states, the magnitude of the orbital angular momentum is equal to $\sqrt{2}\hbar$. For the $p_{\pm1}$ orbitals, the projections along the z axis are $\pm\hbar$. The projections of the angular momentum vectors in all directions in the xy-plane are equally probable, as shown in Figure 6.1. This is qualitatively consistent with the probability distributions of the corresponding electron orbits shown in Figure 6.5(c).

For the p_0 orbital, the orbital angular momentum is also finite and equal to $\sqrt{2}\hbar$, but the projection of the angular momentum vector along the z axis is zero, and the corresponding projection in the xy plane has equal probability of pointing in any direction, as shown in Figure 6.1. Thus, the electron must be orbiting around the nucleus at a distance in planes that contain the z axis but can have any orientation. Thus, qualitatively, the probability distribution function of the electron must be centered on the z axis with azimuthal symmetry in the form shown in Figure 6.5(b).

For the higher excited eigen states of the atom, the wave functions are more complicated spatially. Extending the kind of semi-classical qualitative reasoning given above beyond the simple s and p states to such higher excited states is of questionable validity and usefulness.

6.5 Electron spin and the theory of generalized angular momentum

If an electron is assumed to be a point particle of infinitely small size, it cannot have any spin angular momentum, according to classical mechanics. It is, therefore, not possible to describe the dynamics of the electron spinning motion quantum mechanically on the basis of Schrödinger's representation embodied in Corollary 1 of Postulate 2 (Section 2.2). On the other hand, there are numerous well established experimental observations that could only be understood if an electron has an intrinsic spin. What is implied in the current understanding of the electron is that, although its size is too small to be characterized, its magnetic effects due to its spinning motion are experimentally observable and can be characterized quantum mechanically. Therefore, the quantum theory of atomic particles must be expanded to accommodate the existence of spin angular momentum $\vec{S}$. For this purpose, a generalized angular momentum vector $\vec{J}$, represented by the operator $\hat{J}$, is introduced which includes both the orbital angular momentum and the spin angular momentum.

Since the generalized angular momentum can also represent only the orbital angular momentum, its components must satisfy the same cyclic commutation relations (6.6) as the orbital angular momentum:

$$[\hat{J}_x, \hat{J}_y] = i\hbar\hat{J}_z, \quad [\hat{J}_y, \hat{J}_z] = i\hbar\hat{J}_x, \quad [\hat{J}_z, \hat{J}_x] = i\hbar\hat{J}_y. \tag{6.39}$$

The magnitude of the generalized angular momentum $\hat{J}$ is by definition:

$$\hat{J}^2 \equiv \hat{J}_x^2 + \hat{J}_y^2 + \hat{J}_z^2. \tag{6.40}$$

A right and a left circular component in the xy plane of the generalized angular momentum can also be defined as:

$$\hat{J}_+ = \hat{J}_x + i\hat{J}_y \text{ and } \hat{J}_- = \hat{J}_x - i\hat{J}_y. \tag{6.41}$$

In addition,

$$[\hat{J}^2, \hat{J}_x] = 0, \quad [\hat{J}^2, \hat{J}_y] = 0, \quad [\hat{J}^2, \hat{J}_z] = 0, \tag{6.42}$$

$$[\hat{J}_z, \hat{J}_+] = \hbar \hat{J}_+, [\hat{J}_z, \hat{J}_-] = -\hbar \hat{J}_-, [\hat{J}_+, \hat{J}_-] = 2\hbar \hat{J}_z, \tag{6.43}$$

and

$$[\hat{J}^2, \ \hat{J}_\pm] = 0. \tag{6.44}$$

The cyclic commutation relations (6.39), the sum rules (6.40), and (6.44) can be used as the basis for defining a generalized angular momentum operator $\hat{\vec{J}}$. Any three operators that satisfy the cyclic commutation relations (6.39) can be defined as the three Cartesian components of an equivalent generalized angular momentum vector $\hat{\vec{J}}$. This operator vector gives the angular momentum properties of the particle. This definition certainly applies to the orbital angular momentum operators. In addition, it includes the spin angular momentum of the particle for which there is no Schrödinger's representation of the form (6.4).

At this point, it may not be obvious what has been gained by introducing the concept of the generalized angular momentum beyond what is already known about the orbital angular momentum. The key difference, and a powerful one at that, is that the eigen values of $\hat{\vec{J}}^2$ and $\hat{\vec{J}}_z$ as defined by (6.39) and (6.40) are $j(j+1)\hbar^2$ and $m_j\hbar$, respectively, where j and m_j now include half integers; therefore: $j = 0, 1/2, 1, 3/2, 2, \ldots$ and $m_j = -j, -j+1, -j+2, \ldots j-2, j-1, j$. The formal mathematical proof of this important result is somewhat involved. It would be an unproductive diversion from the discussion here to go into it in detail; it can be found in the literature. (For rigorous mathematical derivations of the properties of the generalized angular momentum, see, for example, Edmonds (1957)). The physical consequences of these additional half-integer eigen values are, however, profound, as we shall see later.

We recall that the results on the eigen values and eigen functions of the orbital angular momentum operators are obtained by solving the corresponding eigen value equations in the Schrödinger representation. For a point particle of infinitely small size, there is no way to write down the spin angular momentum of such a particle according to classical mechanics and make use of the recipe given in Corollary 1 of Postulate 2 (Section 2.2) to arrive at the corresponding eigen value equation in the Schrödinger representation to begin with. On the other hand, there is no problem in writing down the corresponding eigen value equations in the operator form or in Heisenberg's matrix form:

$$\hat{J}^2|jm_j\rangle = j(j+1)\hbar^2|jm_j\rangle, \tag{6.45}$$

$$\hat{J}_z|jm_j\rangle = m_j\hbar|jm_j\rangle. \tag{6.46}$$

In the representation in which $\hat{J}^2$ and $\hat{J}_z$ are diagonal, the matrix representations of the generalized angular momentum operators are, in analogy with (6.29) to (6.31):

$$\langle jm_j|\hat{J}^2|j'm_j'\rangle = j(j+1)\hbar^2\delta_{jj'}\delta_{m_jm_j'}, \tag{6.47}$$

$$\langle jm_j|\hat{J}_z|j'm_j'\rangle = m_j\hbar\delta_{jj'}\delta_{m_jm_j'}, \tag{6.48}$$

and

$$\langle jm_j|\hat{J}_\pm|j'm_j'\rangle = [(j\mp m_j)(j\pm m_j+1)]^{\frac{1}{2}}\hbar\delta_{jj'}\delta_{m_j,(m_j'\pm1)}, \tag{6.49}$$

in the Dirac notation.

When the generalized angular momentum of the point particle represents the spin angular momentum only, the quantum numbers are such that $j = s$ and $m_j = m_s$. The physical implication is that, although the particle size is too small to characterize, its spin angular momentum has, nevertheless, definite measurable properties. The magnitude of the spin angular momentum vector is equal to $\sqrt{s(s+1)}\hbar$ and its projection along the axis of quantization, the z–axis, is $m_s\hbar$. For $s = 1/2$, the magnitude of the spin-1/2 vector is equal to $\frac{\sqrt{3}}{2}\hbar$, and S_z is either up or down: $+\hbar/2$ or $-\hbar/2$. "Spin-1/2" particles are of great fundamental importance, for they include such elementary particles as electrons, protons, neutrons, etc. that are collectively termed "fermions."

The matrix representations of the spin angular momentum operators $\hat{S}_x$, $\hat{S}_y$, $\hat{S}_z$ and $\hat{S}^2$ of the spin-1/2 particle in the representation in which $\hat{S}^2$ and $\hat{S}_z$ are diagonal are:

$$\hat{S}^2 = \begin{pmatrix} \frac{3}{4} & 0 \\ 0 & \frac{3}{4} \end{pmatrix}\hbar^2, \quad \hat{S}_z = \begin{pmatrix} \frac{1}{2} & 0 \\ 0 & -\frac{1}{2} \end{pmatrix}\hbar, \quad \hat{S}_x = \begin{pmatrix} 0 & \frac{1}{2} \\ \frac{1}{2} & 0 \end{pmatrix}\hbar,$$

$$\hat{S}_y = \begin{pmatrix} 0 & -\frac{i}{2} \\ \frac{i}{2} & 0 \end{pmatrix}\hbar. \tag{6.50}$$

The spin matrices are often expressed in terms of the widely used Pauli-matrices $\hat{\sigma}_x, \hat{\sigma}_y, \hat{\sigma}_z$ by splitting off the $\hbar/2$ factor: $\hat{\vec{S}} \equiv \hat{\vec{\sigma}}\dfrac{\hbar}{2}$, where

$$\hat{\sigma}_x \equiv \begin{pmatrix} 0 & 1 \\ 1 & 0 \end{pmatrix}, \quad \hat{\sigma}_y \equiv \begin{pmatrix} 0 & -i \\ i & 0 \end{pmatrix}, \quad \hat{\sigma}_z \equiv \begin{pmatrix} 1 & 0 \\ 0 & -1 \end{pmatrix}, \tag{6.51}$$

and $\hat{S}^2 \equiv \left(\frac{3\hbar^2}{4}\right)\hat{\sigma}^2 \equiv \left(\frac{3\hbar^2}{4}\right)\begin{pmatrix} 1 & 0 \\ 0 & 1 \end{pmatrix}$. $\hat{\sigma}^2$ is a unit matrix. It can be verified easily that all these spin matrices or the Pauli-matrices satisfy the commutation relationships (6.39), (6.40) and (6.42–6.44). The operator representing the spin angular momentum in an arbitrary direction $\vec{n} = \cos\theta_x\,\vec{x} + \cos\theta_y\,\vec{y} + \cos\theta_z\,\vec{z}$ is $\vec{n}\cdot\hat{\vec{S}} = \vec{n}\cdot\hat{\vec{\sigma}}\frac{\hbar}{2}$, where:

$$\vec{n}\cdot\hat{\vec{\sigma}} = \cos\theta_x\hat{\sigma}_x + \cos\theta_y\hat{\sigma}_y + \cos\theta_z\hat{\sigma}_z$$

$$= \begin{pmatrix} \cos\theta_z & \cos\theta_x - i\,\cos\theta_y \\ \cos\theta_x + i\,\cos\theta_y & -\cos\theta_z \end{pmatrix}. \tag{6.52}$$

Note that the eigen values of $\hat{\sigma}_z = 2/\hbar\hat{S}_z$ are simply $+1$ or -1, and the corresponding eigen states $|\,s, m_s\,\rangle$ are: $|1/2, 1/2\rangle = \begin{pmatrix} 1 \\ 0 \end{pmatrix}$ and $|\,1/2, -1/2\,\rangle = \begin{pmatrix} 0 \\ 1 \end{pmatrix}$, corresponding to the spin-up and spin-down states. Diagonalizing the Pauli-matrix $\hat{\sigma}_x = \frac{2}{\hbar}\hat{S}_x$ gives again the eigen values $+1$ and -1. It means that measurement of the x component of the spin angular momentum of the spin-1/2 particle will also yield the values $+\hbar/2$ or $-\hbar/2$. The corresponding normalized eigen states $|\,1/2, m_x = \pm 1/2\rangle$ in the representation in which $\hat{\sigma}^2$ and $\hat{\sigma}_z$ are diagonal are:

$$|\,1/2,\ m_x = +1/2\rangle = \frac{1}{\sqrt{2}}\begin{pmatrix} 1 \\ 1 \end{pmatrix} = \frac{1}{\sqrt{2}}\left[\begin{pmatrix} 1 \\ 0 \end{pmatrix} + \begin{pmatrix} 0 \\ 1 \end{pmatrix}\right]$$

$$= \frac{1}{\sqrt{2}}[\,|1/2,\ 1/2\rangle + |1/2,\ -1/2\rangle] \tag{6.53}$$

and

$$|\,1/2,\ m_x = -1/2\rangle = \frac{1}{\sqrt{2}}[\,|1/2,\ 1/2\rangle - |1/2,\ -1/2\rangle]. \tag{6.54}$$

This shows the interesting result that the spin state that yields the value $+\hbar/2$ or $-\hbar/2$ in the x direction is a coherent superposition of the spin $+\hbar/2$ or $-\hbar/2$ states in the z direction. That is, if measurement of the x–component of the spin gives the value of, for example, either $+\hbar/2$ or $-\hbar/2$, then subsequent measurement of the z component of the same spin system will have an equal probability of getting a $+\hbar/2$ or $-\hbar/2$ value. The same conclusions about eigen values and eigen states hold for the y component of the Pauli-matrix, $\hat{\sigma}_y = \frac{2}{\hbar}\hat{S}_y$ also. These results have important general implications in the studies of, for example, nuclear magnetic or para-magnetic resonance effects.

The attempt to understand the detailed features of the properties of hydrogen and other atoms in the early days of modern physics led to the indisputable conclusions that the electron is a spin-1/2 particle and has a magnetic moment of $-\left(\frac{e\hbar}{2mc}\right)\hat{\vec{\sigma}}$. The constant $\dfrac{e\hbar}{2mc}$ is generally referred to as the "Bohr magneton" and is numerically equal to 0.927×10^{-20}erg/gauss. Furthermore, as shown by Dirac, there are compelling theoretical reasons based on the requirements of the theory of special relativity that the electron must have a magnetic moment of $-\left(\dfrac{e\hbar}{2mc}\right)\hat{\sigma}$ and a spin angular momentum corresponding to $s = 1/2$. These are profoundly important conclusions that can be tested experimentally, and have been verified in numerous experiments without exception. It is difficult to over-emphasize the importance of the consequences of these results to modern science, such as in spectroscopy and magnetism, and to technology and medicine from all the computer and video storage devices to the magnetic resonance imaging applications, just to mention a few.

In the case when the particle has both an orbital angular momentum $\hat{\vec{L}}$ and a spin angular momentum $\hat{\vec{S}}$, the total angular momentum of the particle is the vector sum of the two:

$$\hat{\vec{J}} = \hat{\vec{L}} + \hat{\vec{S}}, \tag{6.55}$$

and its magnitude squared is:

$$\hat{J}^2 = \hat{L}^2 + \hat{S}^2 + (\hat{\vec{L}} \cdot \hat{\vec{S}} + \hat{\vec{S}} \cdot \hat{\vec{L}}). \tag{6.56}$$

Since the orbital and spinning motions involve different degrees of freedom of a particle, $\hat{\vec{L}}$ and $\hat{\vec{S}}$ must commute, or

$$[\hat{\vec{L}}, \hat{\vec{S}}] = 0; \tag{6.57}$$

and

$$\hat{J}^2 = \hat{L}^2 + \hat{S}^2 + 2\hat{\vec{L}} \cdot \hat{\vec{S}}. \tag{6.58}$$

Because of the commutation relations (6.39), it is clear that $\hat{L}_z$ and $\hat{S}_z$ do not commute with $\hat{\vec{L}} \cdot \hat{\vec{S}}$. Thus, the simultaneous eigen states $|\ell, m_\ell; s, m_s\rangle$ cannot also be simultaneous eigen states of $\hat{J}^2$ and $\hat{J}_z$. On the other hand, based on the commutation relations (6.7), (6.42), and (6.44), $\hat{J}^2$ and $\hat{J}_z$ each commute with $\hat{L}^2$ and $\hat{S}^2$. Therefore, the simultaneous eigen states of these four commuting operators $\hat{J}^2, \hat{J}_z, \hat{L}^2$, and $\hat{S}^2$ are the "spin–orbit coupled states" $|j, m_j, \ell, s\rangle$, which must be linear combinations of the uncoupled states $|\ell, m_\ell; s, m_s\rangle$ in the basis in which $\hat{L}^2, \hat{L}_z, \hat{S}^2$, and $\hat{S}_z$ are diagonal:

$$|j, m_j, \ell, s\rangle = \sum_{m_\ell, m_s} \langle \ell, m_\ell; s, m_s | j, m_j, \ell, s \rangle |\ell, m_\ell; s, m_s\rangle. \tag{6.59}$$

The expansion coefficients $\langle \ell, m_\ell; s, m_s | j, m_j, \ell, s \rangle$ are the so-called vector-coupling, or the Clebsch–Gordon, coefficients. They are either tabulated directly or in terms of the related 3-j symbols of Wigner (see, for example, Edmonds (1957). For a specific example of how the vector-coupling coefficients may be calculated on the basis of degenerate perturbation theory, see Problem 9–1 in Chapter 9). The eigen values corresponding to these eigen states are multiples of half-integers and, from (6.55), they are:

$$j = |\ell - s|, \ |\ell - s| + 1, \ldots, \ |\ell + s| - 1, \ |\ell + s|, \tag{6.60}$$

$$m_j = m_\ell + m_s = -j, \ -j + 1, \ -j + 2, \ldots, \ j - 2, \ j - 1, \ j. \tag{6.61}$$

The theory of generalized angular momentum is an elegant frame work for dealing with the spin angular momentum and the interaction of angular momentum vectors in atomic particles. A specific example is the spin–orbit interaction in the hydrogen atom in particular and in all atoms in general, as will be discussed in the following section.

6.6 Spin–orbit interaction in the hydrogen atom

Since the Hamiltonian including only the Coulomb interaction term is independent of $\hat{S}^2$ or $\hat{S}_z$, the eigen energy levels are degenerate with respect to the spin angular moment quantum numbers s and m_s. It also means that there are two more constants of motion in addition to those associated with the orbital angular momentum: the magnitude and the z component of the spin angular momentum vectors. Thus, taking the electron spin into account, there are now five good quantum numbers to completely specify the stationary states of the hydrogen atom: $|n, \ell, m_\ell, s, m_s\rangle$, because there are five commuting operators: $\hat{H}, \ \hat{L}^2, \ \hat{L}_z, \ \hat{S}^2, \ \hat{S}_z$.

Taking spin into account, there is actually a magnetic interaction between the orbital motion and the spinning motion of the electron. Qualitatively, the physical origin of the interaction is that the orbital motion of a charged particle such as the electron sets up a magnetic field $\vec{B}$ which is proportional to the orbital angular momentum $\vec{L}$. It is a relativistic effect: in the rest-frame of the electron moving with the velocity $\vec{v}$ in the Coulomb potential $V(r)$ (here, V refers to the electrical potential, not the electron potential-energy as defined elsewhere in this book) of the nucleus with the charge $+Ze$, it sees an effective magnetic field:

$$\vec{B}' = -\frac{1}{c}\nabla V(r) \times \vec{v} = \left(\frac{Ze}{c}\right)\frac{\vec{r}}{r^3} \times \vec{v} = \left(\frac{Ze}{mcr^3}\right)\vec{L}.$$

The magnetization associated with the spinning motion of the electron is proportional to the spin angular momentum $\vec{S}$ and is equal to $-\dfrac{e}{mc}\vec{S}$. The corresponding magnetic interaction energy $-\vec{B}' \cdot \vec{M}$ is then proportional to the product of the orbital angular momentum and the spin angular momentum: $\vec{L} \cdot \vec{S}$. Taking into account the Thomas precession effect, which requires an extra factor of $(1/2)$ (see, for example, Bethe and Jackiew (1986), p.152), it leads to a "spin–orbit interaction" term in the Hamiltonian of the form:

$$\hat{H}_{s-o} = \left(\frac{Ze^2}{2m^2c^2r^3}\right)\hat{\vec{L}}\cdot\hat{\vec{S}} \equiv \zeta(r)\hat{\vec{L}}\cdot\hat{\vec{S}}. \tag{6.62}$$

The proportionality constant $\zeta(r)$ is generally known as the "spin–orbit parameter." It is a function of the distance r of the electron from the nucleus, the positive charge Ze in the nucleus, and the mass of the orbiting electron. This model can be extended to describe the spin–orbit interactions in "hydrogenic atoms or ions" (with $Z > 1$) or charged particles in solids in general. In that case, $\zeta(r)$ may be a more complicated function of r than that shown in (6.62) and the electron mass may be replaced by an "effective mass" of the charged particle in the solid.

The total Hamiltonian of the hydrogen atom ($Z = 1$) including the spin–orbit interaction is, therefore:

$$\hat{H} = -\frac{\hbar^2}{2m}\nabla^2 - \frac{e^2}{r} + \zeta(r)\,\hat{\vec{L}}\cdot\hat{\vec{S}} = -\frac{\hbar^2}{2m}\nabla^2 - \frac{e^2}{r} + \frac{\zeta(r)}{2}\,(\hat{J}^2 - \hat{L}^2 - \hat{S}^2). \tag{6.63}$$

Note that this Hamiltonian no longer commutes with $\hat{L}_z$ and $\hat{S}_z$, but with $\hat{J}^2$, $\hat{J}_z$, $\hat{L}^2$, and $\hat{S}^2$. The z components of the orbital and the spin angular momentum vectors are no longer constants of motion. Instead, the magnitude and the z component of the spin–orbit coupled total angular momentum vector, $\hat{J}^2$ and $\hat{J}_z$, are the new constants of motion in addition to the magnitude of the orbital and the spin angular momentum. Thus, taking the spin–orbit interaction into account, the stationary eigen states of the hydrogen atom are the simultaneous eigen states $|n, j, m_j, \ell, s\rangle$ of the Hamiltonian and these four commuting operators ($\hat{J}^2$, $\hat{J}_z$, $\hat{L}^2$, and $\hat{S}^2$). The energy eigen values, $E_{n\ell j}$, are no longer as simple as that shown in Eq. (6.37). They depend not only on the principal quantum number n, but also on the orbital and the total angular quantum numbers ℓ and j, respectively. The radial wave functions are also considerably more complicated because the spin–orbit parameter $\zeta(r)$ is a function of r and has to be evaluated numerically. The angular and the spin part of the wave functions are the spin–orbit coupled states $|j, m_j, \ell, s\rangle$ of the form given in (6.59). Assuming, however, that the effect of the spin–orbit interaction on the radial wave function is much smaller than that of the rest of the terms in the Hamiltonian and is negligible, the eigen values $E_{nj\ell}$ become approximately, from (6.63):

$$E_{n\ell j} \cong E_n + \frac{\zeta_{n\ell}\hbar^2}{2}\left[j(j+1) - \ell(\ell+1) - \frac{1}{2}\cdot\frac{3}{2}\right], \tag{6.64}$$

where $\zeta_{n\ell} \approx \int_0^\infty \zeta(r)|R_{n\ell}(r)|^2 r^2 \mathrm{d}r$. It shows that the manifold of degenerate states with the same ℓ and s values but different m_ℓ and m_s values are now split into two groups of still degenerate states with shifts of $\zeta_{n\ell}\hbar^2\ell/2$ and $-\zeta_{n\ell}(\ell+1)\hbar^2/2$ corresponding to $j = \ell + 1/2$ and $j = \ell - 1/2$, respectively. Thus, the accidental degeneracy of the hydrogen atom is now lifted, because of the ℓ and j dependence in the energy eigen values, as anticipated earlier in Section 6.4. For example, the 2p level is now split into two groups corresponding to two degenerate levels with shifts of $\zeta_{n\ell}\hbar^2/2$ and $-\zeta_{n\ell}\hbar^2$ corresponding to $j = 3/2$ ($m_j = \pm 3/2, \pm 1/2$) and $j = 1/2$ ($m_j = \pm 1/2$), respectively. This splitting is called the "spin–orbit splitting." The spin–orbit parameter $\zeta_{n\ell}$ can be regarded as a fitting parameter that can be determined from experimental data on such a splitting.

6.7 Problems

6.1 Give the matrix representations of the angular momentum operators $\hat{L}_x$, $\hat{L}_y$, $\hat{L}_z$, $\hat{L}_+$, $\hat{L}_-$, and $\hat{L}^2$, for $\ell = 0, 1$, and 2, in the basis in which $\hat{L}_z$ and $\hat{L}^2$ are diagonal.

6.2 Using the matrix representations of the Cartesian components of the angular momentum operators for $\ell = 1$ and 2 found in Problem 6.1, show that the cyclic commutation relationships (6.6) and (6.7) are indeed satisfied.

6.3 Since the three Cartesian components of the orbitial angular momentum operators do not commute with each other, does that mean we can never specify the three components of the orbitial angular momentum of hydrogen atom precisely simultaneously? If that is not the case, give the conditions when they can and cannot be specified simultaneously and why.

6.4 Show that the $n = 2$, $\ell = 1$, and $m_\ell = 1$ wave function indeed satisfies the time-independent Schrödinger equation given in the text for the hydrogen atom. Show explicitly also that this wave function is normalized.

6.5 A particle is known to be in a state such that $\hat{L}^2 = 2\hbar^2$. It is also known that measurement of $\hat{L}_z$ will yield the value $+\hbar$ with the probability $1/3$ and the value $-\hbar$ with the probability $2/3$.

(a) What is the normalized wave function, $\Psi(\theta, \phi)$, of this particle in terms of the spherical harmonics?

(b) What is the expectation value, $\langle \hat{L}_z \rangle$, of the z component of the angular momentum of this particle?

6.6 The wave function of a particle of mass m moving in a potential well is, at a particular time t:

$$\Psi(x, y, z) = (x + y + z)\, e^{-\alpha\sqrt{x^2+y^2+z^2}}.$$

(a) Write Ψ in the spherical coordinate system and normalize the wave function, $\Psi(\theta, \phi)$.

(b) What is the probability measurement of $\hat{L}^2$ and $\hat{L}_z$ gives the values $2\hbar^2$ and 0, respectively?

6.7 Consider a mixed state of hydrogen:

$$\Psi = R_{21}(r)\, Y_{11}(\theta, \phi) + 2R_{32}(r)\, Y_{21}(\theta, \phi).$$

(a) Normalize Ψ.

(b) Is Ψ an eigen function of $\hat{L}^2$; of $\hat{L}_z$? Explain.

(c) Calculate the expectation value $\langle \Psi | \hat{L}^2 | \Psi \rangle$ in terms of $\hbar$.

(d) Calculate $\langle \Psi | \hat{L}_z | \Psi \rangle$ in terms of $\hbar$.

(e) Calculate $\langle \Psi | \hat{H} | \Psi \rangle$. Give your answer in eV.

6.8 Consider a hydrogen atom in the following mixed state at $t = 0$:

$$\Psi(r, \theta, \phi, t = 0) = 3R_{32}(r) Y_{20}(\theta, \phi) + R_{21}(r) Y_{11}(\theta, \phi).$$

(a) Normalize the wave function.
(b) Is the atom in a stationary state? Explain briefly.
(c) What is the expectation value of the energy at $t > 0$?
(d) What is the expectation value of $\hat{L}^2$ and $\hat{L}_z$ at $t = 0$?
(e) What is the uncertainty of $\hat{L}_z$ in this state?

$$\Delta L_z = [\langle \Psi | \hat{L}_z | \Psi \rangle - \langle \Psi | \hat{L}_z | \Psi \rangle^2]^{\frac{1}{2}}.$$

6.9 A particle of mass m is placed in a finite spherical well:

$$V(r) = \begin{cases} 0, & \text{if } r \leq a, \\ V_0, & \text{if } r \geq a. \end{cases}$$

Find the ground state wave function by solving the radial equation with $\ell = 0$. Show that there is no bound state at all if $V_0 a^2 < \pi^2 \hbar^2 / 8m$.

7 Multi-electron ions and the periodic table

An electron in a hydrogenic atom or ion can occupy any of the $|n\ell s j m_j\rangle$ eigen states of the Hamiltonian of the atom or ion. In ions or atoms with more than one electron, the solutions of the time independent Schrödinger equations become complicated because the electrons interact not only with the positively charged nucleus, but also with each other. Particles with half-integer spin angular momentum, such as electrons, must also satisfy Pauli's exclusion principle, which forbids two such particles to occupy the same quantum state. Furthermore, the electrons in the multi-electron ion or atom are indistinguishable from one another. Taking these considerations into account, the electrons will systematically fill all the available single-electron states of successively higher energies in multi-electron ions or atoms. Because of the nature of the quantum states occupied by the electrons, the physical and chemical properties of the elements exhibit certain patterns and trends which form the basis of the periodic table.

7.1 Hamiltonian of the multi-electron ions and atoms

Consider an ion with N electrons and Z protons in the nucleus; for a neutral multi-electron atom, $Z = N$. Again, because the nucleus is much heavier than the electrons, we assume it to be stationary at the origin of a spherical coordinate system, as shown in Figure 7.1.

Including only the kinetic energy of the electrons and the potential energy due to the electrostatic interactions among the electrons and between the electrons and the nucleus, the Hamiltonian of the electrons in the ion for the orbital part of the motion only is:

$$\hat{H} = \sum_{i=1}^{N} [-\frac{\hbar^2}{2m}\nabla_i^2 - \frac{Ze^2}{r_i}] + \sum_{i>j=1}^{N} \frac{e^2}{r_{ij}}. \tag{7.1}$$

The form of the summation sign in the last term is to ensure that the electrostatic interaction between each pair of electrons is counted only once. The factor

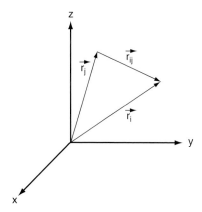

Figure 7.1. Coordinate system used for the model for the multi-electron ion or atom. The nucleus is assumed stationary at the origin $(0, 0, 0)$.

$r_{ij} = \sqrt{(x_i - x_j)^2 + (y_i - y_j)^2 + (z_i - z_j)^2}$ in the last term of (7.1) makes it impossible to solve the equation without approximations.

The standard approximation procedure is to assume that each electron is moving primarily in a spherically symmetric potential $V(r)$, due to the nucleus and the average potential of all the other electrons:

$$\hat{H} = \sum_{i=1}^{N} [-\frac{\hbar^2}{2m}\nabla_i^2 + V(r_i)] + \Delta V_{es}, \tag{7.2}$$

where

$$\Delta V_{es} = \left\{ \sum_{i>j=1}^{N} \frac{e^2}{r_{ij}} - \sum_{i=1}^{N} \left[\frac{Ze^2}{r_i} + V(r_i) \right] \right\} \approx 0 \tag{7.3}$$

is considered a negligibly small perturbation in a first order approximation; or:

$$\hat{H} \cong \left\{ \sum_{i=1}^{N} [-\frac{\hbar^2}{2m}\nabla_i^2 + V(r_i)] \right\}. \tag{7.4}$$

Thus, the time-independent Schrödinger equation for the multi-electron ion or atom to be solved approximately is:

$$\left\{ \sum_{i=1}^{N} [-\frac{\hbar^2}{2m}\nabla_i^2 + V(r_i)] \right\} \Psi_{\{E\}_n}(\vec{r}_1, \vec{r}_2, \ldots, \vec{r}_{N-1}, \vec{r}_N)$$

$$= \{E\}_n \Psi_{\{E\}_n}(\vec{r}_1, \vec{r}_2, \ldots, \vec{r}_{N-1}, \vec{r}_N). \tag{7.5}$$

7.2 Solutions of the time-independent Schrödinger equation for multi-electron ions and atoms

Because the differential operator in Eq. (7.5) is comprised of separate terms of the electron coordinates $\vec{r}_i$ of the same form, its eigen functions must be products of the eigen functions of the individual differential-operator terms:

$$\Psi_{\{E\}_n}(\vec{r}_1, \vec{r}_2, \dots, \vec{r}_{N-1}, \vec{r}_N) = \prod_{i=1}^{N} \Psi_{E_i}(\vec{r}_i)$$

$$= \Psi_{E_1}(\vec{r}_1)\Psi_{E_2}(\vec{r}_2) \dots \Psi_{E_{N-1}}(\vec{r}_{N-1})\Psi_{E_N}(\vec{r}_N), \tag{7.6}$$

where

$$\left[-\frac{\hbar^2}{2m}\nabla_i^2 + V(r_i)\right]\Psi_{E_i}(\vec{r}_i) = E_i\Psi_{E_i}(\vec{r}_i), \tag{7.7}$$

and the eigen value $\{E\}_n$ is the total energy of the atom and must be the sum of the individual eigen values E_i:

$$\{E\}_n = E_1 + E_2 + \dots + E_{N-1} + E_N \equiv \sum_{i=1}^{N} E_i. \tag{7.8}$$

Thus, the key to solving Eq. (7.5) is to find the eigen states and eigen values of the single-electron Hamiltonian, (7.7). The only difference between this equation and the time-independent Schrödinger equation for the hydrogen atom given in Chapter 6 is in the form of the spherical potential term $V(r)$, which is a function of r only, but not of θ and ϕ, in the present approximation. Therefore, the single-electron orbital states are:

$$\Psi_{n_i\ell_i m_{\ell_i}}(r_i, \theta_i, \phi_i) = R_{n_i\ell_i}(r_i)Y_{\ell_i m_{\ell_i}}(\theta_i, \phi_i), \tag{7.9}$$

where, from (6.36), the radial part of the wave function satisfies the equation:

$$\left\{-\frac{\hbar^2}{2m}\left[\frac{1}{r^2}\frac{\partial}{\partial r}\left(r^2\frac{\partial}{\partial r}\right) - \frac{\ell(\ell+1)}{r^2}\right] + V(r)\right\}R_{E_\ell}(r) = E_\ell R_{E_\ell}(r), \tag{7.10}$$

and the angular part is the known spherical harmonics $Y_{\ell m_\ell}(\theta, \phi)$. The $|n_i\ell_i m_{\ell_i}\rangle$ states, are the available single-electron states of the multi-electron ion or atom. The N electrons of the multi-electron ion or atom will occupy some of these available single-electron states, and the resulting wave functions for the ion or atom are basically of the form (7.6). There are, however, other considerations, such as the permutation degeneracy, the indistinguishability of the electrons, the Pauli exclusion principle, and the electron spin that must be taken into account, as will be discussed later. Taking these into account will make the wave functions of the multi-electron ions or atoms much more complicated.

The angular part of the wave function $\Psi_{n_i \ell_i m_{\ell_i}}(r_i, \theta_i, \phi_i)$ is not a problem; it is the spherical harmonics. To find the radial part of the solution of (7.10), the spherical potential $V(r_i)$ must be specified. Since it represents the average potential due to the effects of the nucleus and all the other electrons on the i-th electron, rigorously speaking, this means that we must know the wave functions of all the electrons before we can specify $V(r)$ for any one of the electrons. This then becomes a circular problem and an impossible task, since this would require solving the equations of the form (7.10) for all the electrons simultaneously before knowing what $V(r)$ is. Looking for rigorous eigen functions and eigen values of the exact Hamiltonian (7.1) is, therefore, not what is done in practice. Fortunately, to understand the properties of the multi-electron ion or atom, it is not necessary in general to know the detailed form of the radial part of the wave functions. It is sufficient to know that, in most cases, the radial wave functions can be derived using for $V(r)$ a "shielded Coulomb potential energy" model in the Hamiltonian; that is, near the nucleus, it can be closely approximated by a Coulomb potential, and outside of some nominal distance α^{-1} (called the "Debye length") from the nucleus the potential becomes exponentially smaller with distance (known as the "screened-Coulomb" or "Yukawa" potential):

$$V(r) \approx -\frac{e^2}{r} \exp(-\alpha r).$$

Numerically, there are well-developed iterative procedures, such as the Hartree or Hartree–Fock method, for calculating the eigen functions and eigen values of the more exact Hamiltonian. (See, for example, Bethe and Jackiw (1986)). It is beyond the scope of the present discussion to get involved in such topics.

To proceed, we will assume that the single-electron wave function $\Psi_{n_i \ell_i m_{\ell_i}}(r_i, \theta_i, \phi_i)$ and the corresponding eigen value $E_{n_i \ell_i m_{\ell_i}}$ can be obtained by one method or another and are known. The eigen functions and the corresponding eigen values of the multi-electron ion or atom will then depend on which of these single-electron states are occupied, subject to the following considerations:

1. For more than one electron, there is an additional degeneracy called "permutation" degeneracy, meaning the assignment of the electrons to the occupied single-electron states is not unique and can be permuted. This degeneracy will ultimately be removed, however, when the following consideration (2 below) is taken into account.
2. Since "all electrons are alike," they cannot be distinguished from one another. That is, the "indistinguishability" of the electrons occupying the same general space in the atom or ion must be taken into account.
3. There is the additional basic postulate of Pauli's exclusion principle, which says that no two "fermions" (particles of half-integer spin angular momentum quantum numbers), such as electrons, can occupy the same quantum state defined over the same space.
4. Given the indistinguishability of the electrons, the wave function squared must not change when the coordinates of any two particles in the atomic wave function are exchanged. This means that the atomic wave function itself must either remain invariant or, at most, change sign upon exchanging the coordinates of any two

particles. It is known from experiment that for fermions, and hence electrons, the atomic wave function changes sign. Pauli's exclusion principle then necessarily follows as a consequence for fermions. For bosons (particles of integral spin angular momentum quantum numbers, such as photons), the wave function does not change sign and remains invariant; therefore, Pauli's exclusion principle does not apply to bosons.

Let us now see how these new considerations will ultimately determine the form of the wave functions for the energy eigen states of the multi-electron ion or atom.

First, in the product wave function of the form, (7.6), each electron ($\vec{r}_i$) is uniquely associated with one specific quantum state ($n_i \ell_i m_{\ell_i}$):

$$\Psi_{\{E\}_n}(\vec{r}_1, \vec{r}_2, \ldots, \vec{r}_{N-1}, \vec{r}_N)$$

$$= \Psi_{n_1 \ell_1 m_{\ell_1}}(\vec{r}_1) \Psi_{n_2 \ell_2 m_{\ell_2}}(\vec{r}_2) \ldots \Psi_{n_{N-1} \ell_{N-1} m_{\ell_{N-1}}}(\vec{r}_{N-1}) \Psi_{n_N \ell_N m_{\ell_N}}(\vec{r}_N).$$

But, since the electrons in the multi-electron atom are indistinguishable from one another, the i th electron may just as well be in the j th eigen state. For example: electron 1 could be in the single-electron state specified by the set of quantum numbers a_2, and electron 2 could well be in the state a_1. (To simplify the notations, instead of spelling out the quantum numbers $n_i \ell_i m_{\ell_i}$ repeatedly, we have used the symbol a_i to represent the whole set of good quantum numbers.) Thus, the corresponding multi-electron wave function:

$$\Psi_{a_1 a_2 \ldots a_{N-1} a_N}(\vec{r}_1, \vec{r}_2, \ldots, \vec{r}_{N-1}, \vec{r}_N)$$

$$= \Psi_{a_2}(\vec{r}_1) \Psi_{a_1}(\vec{r}_2) \ldots \Psi_{n_{N-1} \ell_{N-1} m_{\ell_{N-1}}}(\vec{r}_{N-1}) \Psi_{n_N \ell_N m_{\ell_N}}(\vec{r}_N)$$

is also a valid eigen state of the Hamiltonian corresponding to the energy $\{E\}_n$. In fact any electron can be associated with any one of the occupied eigen states, or any of the multi-electron wave functions with the electrons permuted among the occupied single-electron states is a valid eigen functions corresponding to the same eigen value $\{E_n\}$ with equal probability. Such a degeneracy is called "permutation degeneracy" for the multi-electron ion or atom. This degeneracy will, however, be eliminated, when the indistinguishability of the electrons in the multi-electron ion or atom (consideration 2 above) is taken into account. The true eigen state of the multi-electron atom is then the superposition state constructed from all these permuted degenerate states with equal probability. It turns out that there is an elegant form of such a state function that not only satisfies that requirement but also Pauli's exclusion principle (consideration 3 above). It is known as the "Slater determinant":

$$\Psi_{a_1 a_2 \ldots a_{N-1} a_N}(\vec{r}_1, \vec{r}_2, \ldots, \vec{r}_{N-1}, \vec{r}_N)$$

$$= \frac{1}{\sqrt{N!}} \begin{vmatrix} \Psi_{a_1}(\vec{r}_1) & \Psi_{a_1}(\vec{r}_2) & \ldots & \Psi_{a_1}(\vec{r}_{N-1}) & \Psi_{a_1}(\vec{r}_N) \\ \Psi_{a_2}(\vec{r}_1) & \Psi_{a_2}(\vec{r}_2) & \ldots & \Psi_{a_2}(\vec{r}_{N-1}) & \Psi_{a_2}(\vec{r}_N) \\ \ldots & \ldots & \ldots & \ldots & \ldots \\ \ldots & \ldots & \ldots & \ldots & \ldots \\ \ldots & \ldots & \ldots & \ldots & \ldots \\ \Psi_{a_{N-1}}(\vec{r}_1) & \Psi_{a_{N-1}}(\vec{r}_2) & \ldots & \Psi_{a_{N-1}}(\vec{r}_{N-1}) & \Psi_{a_{N-1}}(\vec{r}_N) \\ \Psi_{a_N}(\vec{r}_1) & \Psi_{a_N}(\vec{r}_2) & \ldots & \Psi_{a_N}(\vec{r}_{N-1}) & \Psi_{a_N}(\vec{r}_N) \end{vmatrix}. \tag{7.11}$$

Note that every possible set of the electron coordinates $\vec{r}_i$ is associated with every possible set of quantum numbers a_i in all possible permutations with equal probability, thus satisfying consideration 2 above. Furthermore, if any a_i is the same as any a_j, the Slater-determinant (7.11) would automatically vanish, thus satisfying consideration 3 above. Finally, exchanging the coordinates of any two particles, $\vec{r}_i \leftrightarrow \vec{r}_j$, is equivalent to exchanging two columns in the Slater-determinant in (7.11). It changes the sign of the determinant above, as required of fermions (consideration 4).

Also, to include the spin of the electrons, we can expand the definition of a_i to include the spin angular momentum as well; thus, a_i represents the complete set of five good quantum numbers $(n_i, \ell_i, m_{\ell_i}, s = 1/2, m_{s_i})$. If the individual spin–orbit interaction of each electron is to be taken into account, the quantum numbers can simply be replaced by $(n_i, \ell_i, s = 1/2, j_i, m_{j_i})$. We have then, in principle, the approximate wave function of any state of any multi-electron ion or atom. Depending on how the neglected higher order perturbation terms are taken into account, the improved eigen functions of the original Hamiltonian, (7.1), will be various linear combinations of such multi-electron wave functions. For real ions or atoms, however, the simple picture presented here is only a good model for a qualitative understanding of the general properties of the ions and atoms with 'not too many electrons.' Even in the "simple" cases, calculation of the radial wave functions is not a simple matter. Nevertheless, it is amazing how much can be learned from such a simple model, as we shall see.

7.3   The periodic table

For a multi-electron atom in the ground state, the electrons will fill the available single-electron states one by one from the lowest energy states, the 1s states, up. The 1s level has no orbital degeneracy but a spin degeneracy of 2. Starting with the one-electron atom, the single electron in the hydrogen atom is in the $n = 1$, $\ell = 0$, and $m_\ell = 0$ orbital state, but it can be in either of the spin degenerate states with $s = 1/2$ and $m_s = \pm 1/2$. In the hydrogen atom, since only one of the two available spin states of the 1s level is filled, there is a tendency for the atom to accommodate another electron of the opposite spin from another atom and form a molecule. This kind of bonding of two atoms to form a molecule by sharing electrons is called "covalent bonding," as we shall see later.

Next, in the two-electron atom, helium, the two electrons must be in the $n = 1$, $\ell = 0$, $m_\ell = 0$, $s = 1/2$ and $m_s = \pm 1/2$ states. It is interesting to note that, once the two available single-electron states in the 1s level are filled, it is less likely for the atom to bond with other atoms to form a molecule, and the atom becomes relatively "inert" chemically. The detailed reason is more involved. Qualitatively, it is because the next available single-electron state in the helium atom is far above the ground state in energy. Therefore, it is not energetically favorable for electrons from another atom to be near the nucleus and the electrons of a helium atom in a stable molecular configuration. For the same reason, neon (10 electrons, all the $n = 1$ and $n = 2$ states are filled) and argon (18 electrons, all states up to the 3p and 3s states) are also inert gases. If spin–orbit interaction is taken into account, then it is the $|n, \ell, s, j, m_j\rangle$ states, not the $|n, \ell, m_\ell, s, m_s\rangle$ states, that must be filled successively.

As the number of electrons in the atom increases, they will fill states of successively higher energy. As long as the number of electrons is not too large, the pattern of the few occupied eigen states of the atoms are similar to that of the hydrogenic states, as shown in Figure 6.3. Thus, the electrons tend to fill the states with smaller principal quantum numbers first, forming "filled shells." Within each manifold of states with the same principal quantum number, the s and p states tend to get filled first, but when there are more and more electrons so that the d states are beginning to be occupied, the pattern tends to become less and less clear, because the hydrogenic model of the multi-electron atom is less and less realistic for such multi-electron ions or atoms. In such cases, the states with the same principal quantum numbers may not all be filled sequentially. When the available s and p states are filled, they tend to be inert chemically, however. Examples are krypton (36 electrons, all the ns^2p^6 states up to $n = 4$ are filled, but not the 4d states), xenon (54 electrons, all the ns^2p^6 states up to $n = 5$ are filled, but not the 5d and 5f states), etc. They are all inert gases.

If the chemical elements are tabulated according to the types of the orbitals of the "valence electrons," or the electrons in the outermost shells, we end up with what is known as the "periodic table." For the purpose of illustration, the first few rows of the periodic table involving elements with valence electrons with principal quantum numbers up to $n = 6$ are shown in Table 7.1. The elements in each row are arranged in order according to the total number of electrons in the elements. For the first four rows, the configurations given refer to the valence electrons only; the designations of the electrons in the closed shells are suppressed. For example, a neutral gallium (Ga) atom has a total of 31 electrons. Its full ground state configuration is: $(1s)^2(2s)^2(2p)^6(3s)^2(3p)^6(3d)^{10}(4s)^2(4p)$. Only $(4s)^2(4p)$ is shown. For elements with valence electrons with n greater than 5, some of the ns and np states are occupied before all the available d and f states with lower n values are occupied due to energy considerations. The columns that are labeled from I to VIII refer to elements with s- and p-electrons in the valence shells. For the ones that are not labeled, the outer-most d- and f-electrons are also shown. A fuller table can be found in many introductory physics or chemistry text books and will not be repeated here (see, for example, Kittel (1996)).

It was known long before the development of quantum mechanics that if the elements were arranged more or less as in the periodic table, there were certain similarities between the chemical properties of the elements of the same column of the first few rows, and certain trends from element to element of the same row. With the development of quantum mechanics, such patterns and trends can be understood qualitatively on the basis of the nature of the wave functions of the valence electrons. The elements of the same column have valence orbitals of the same type. For example, the first few column-IV elements: carbon, silicon, and germanium, all have four valence electrons with s^2p^2 orbitals. The geometry of these orbitals are similar, as shown in Figure 6.5. The crystalline solids formed from these atoms tend to have the same structure and similar electronic properties, because of the nature and geometry

Table 7.1. *Partial Periodic Table.*

Main groups

I	II	III	IV	V	VI	VII	VIII
$\mathbf{H}^{1}$ $1s$							$\mathbf{He}^{2}$ $1s^2$
$\mathbf{Li}^{3}$ $2s$	$\mathbf{Be}^{4}$ $2s^2$	$\mathbf{B}^{5}$ $2p$ $2s^2$	$\mathbf{C}^{6}$ $2p^2$ $2s^2$	$\mathbf{N}^{7}$ $2p^3$ $2s^2$	$\mathbf{O}^{8}$ $2p^4$ $2s^2$	$\mathbf{F}^{9}$ $2p^5$ $2s^2$	$\mathbf{Ne}^{10}$ $2p^6$ $2s^2$
$\mathbf{Na}^{11}$ $3s$	$\mathbf{Mg}^{12}$ $3s^2$	$\mathbf{Al}^{13}$ $3p$ $3s^2$	$\mathbf{Si}^{14}$ $3p^2$ $3s^2$	$\mathbf{P}^{15}$ $3p^3$ $3s^2$	$\mathbf{S}^{16}$ $3p^4$ $3s^2$	$\mathbf{Cl}^{17}$ $3p^5$ $3s^2$	$\mathbf{Ar}^{18}$ $3p^6$ $3s^2$
$\mathbf{K}^{19}$ $4s$	$\mathbf{Ca}^{20}$ $4s^2$	$\mathbf{Ga}^{31}$ $4p$ $4s^2$	$\mathbf{Ge}^{32}$ $4p^2$ $4s^2$	$\mathbf{As}^{33}$ $4p^3$ $4s^2$	$\mathbf{Se}^{34}$ $4p^4$ $4s^2$	$\mathbf{Br}^{35}$ $4p^5$ $4s^2$	$\mathbf{Kr}^{36}$ $4p^6$ $4s^2$
$\mathbf{Rb}^{37}$ $5s$	$\mathbf{Sr}^{38}$ $5s^2$	$\mathbf{In}^{49}$ $5p$ $5s^2$	$\mathbf{Sn}^{50}$ $5p^2$ $5s^2$	$\mathbf{Sb}^{51}$ $5p^3$ $5s^2$	$\mathbf{Te}^{52}$ $5p^4$ $5s^2$	$\mathbf{I}^{53}$ $5p^5$ $5s^2$	$\mathbf{Xe}^{54}$ $5p^6$ $5s^2$
$\mathbf{Cs}^{55}$ $6s$	$\mathbf{Ba}^{56}$ $6s^2$	$\mathbf{Tl}^{81}$ $6p$ $6s^2$	$\mathbf{Pb}^{82}$ $6p^2$ $6s^2$	$\mathbf{Bi}^{83}$ $6p^3$ $6s^2$	$\mathbf{Po}^{84}$ $6p^4$ $6s^2$	$\mathbf{At}^{85}$ $6p^5$ $6s^2$	$\mathbf{Rn}^{86}$ $6p^6$ $6s^2$

Transition metals (d-block)

$\mathbf{Sc}^{21}$ $3d$ $4s^2$	$\mathbf{Ti}^{22}$ $3d^2$ $4s^2$	$\mathbf{V}^{23}$ $3d^3$ $4s^2$	$\mathbf{Cr}^{24}$ $3d^5$ $4s$	$\mathbf{Mn}^{25}$ $3d^5$ $4s^2$	$\mathbf{Fe}^{26}$ $3d^6$ $4s^2$	$\mathbf{Co}^{27}$ $3d^7$ $4s^2$	$\mathbf{Ni}^{28}$ $3d^8$ $4s^2$	$\mathbf{Cu}^{29}$ $3d^{10}$ $4s$	$\mathbf{Zn}^{30}$ $3d^{10}$ $4s^2$
$\mathbf{Y}^{39}$ $4d$ $5s^2$	$\mathbf{Zr}^{40}$ $4d^2$ $5s^2$	$\mathbf{Nb}^{41}$ $4d^4$ $5s$	$\mathbf{Mo}^{42}$ $4d^5$ $5s$	$\mathbf{Tc}^{43}$ $4d^6$ $5s$	$\mathbf{Ru}^{44}$ $4d^7$ $5s$	$\mathbf{Rh}^{45}$ $4d^8$ $5s$	$\mathbf{Pd}^{46}$ $4d^{10}$ $-$	$\mathbf{Ag}^{47}$ $4d^{10}$ $5s$	$\mathbf{Cd}^{48}$ $4d^{10}$ $5s^2$
$\mathbf{La}^{57}$ $5d$ $6s^2$	$\mathbf{Hf}^{72}$ $4f^{14}$ $5d^2$ $6s^2$	$\mathbf{Ta}^{73}$ $4f^{14}$ $5d^3$ $6s^2$	$\mathbf{W}^{74}$ $4f^{14}$ $5d^4$ $6s^2$	$\mathbf{Re}^{75}$ $4f^{14}$ $5d^5$ $6s^2$	$\mathbf{Os}^{76}$ $4f^{14}$ $5d^6$ $6s^2$	$\mathbf{Ir}^{77}$ $4f^{14}$ $5d^9$ $-$	$\mathbf{Pt}^{78}$ $4f^{14}$ $5d^9$ $6s$	$\mathbf{Au}^{79}$ $4f^{14}$ $5d^{10}$ $6s$	$\mathbf{Hg}^{80}$ $4f^{14}$ $5d^{10}$ $6s^2$

f-block (lanthanides)

$\mathbf{Ce}^{58}$ $4f^2$ $6s^2$	$\mathbf{Pr}^{59}$ $4f^3$ $6s^2$	$\mathbf{Nd}^{60}$ $4f^4$ $6s^2$	$\mathbf{Pm}^{61}$ $4f^5$ $6s^2$	$\mathbf{Sm}^{62}$ $4f^6$ $6s^2$	$\mathbf{Eu}^{63}$ $4f^7$ $6s^2$	$\mathbf{Gd}^{64}$ $4f^7$ $5d$ $6s^2$	$\mathbf{Tb}^{65}$ $4f^8$ $5d$ $6s^2$
$\mathbf{Dy}^{66}$ $4f^{10}$ $6s^2$	$\mathbf{Ho}^{67}$ $4f^{11}$ $6s^2$	$\mathbf{Er}^{68}$ $4f^{12}$ $6s^2$	$\mathbf{Tm}^{69}$ $4f^{13}$ $6s^2$	$\mathbf{Yb}^{70}$ $4f^{14}$ $6s^2$	$\mathbf{Lu}^{71}$ $4f^{14}$ $5d$ $6s^2$		

of these orbitals. All the rest of the electrons in these atoms have smaller orbits and are more tightly bound to the nucleus than, and are shielded by, the valence electrons. It is the valence electrons of an atom that tend to respond more readily to any external perturbations, such as an applied electric field or in chemical reactions, and determine, for example, the optical and chemical properties of the element. Also, from the outside world, it is the geometry of these valence orbitals that determines the "shape" of the atom, and thus the structure of the molecules and crystalline solids formed from such atoms, as will be shown in later chapters.

7.4 Problems

7.1. Show that the Slater determinant for a two-electron atom in the form given in (7.11) is normalized, if all the single-electron wave functions in the determinant are normalized.

7.2. Write out the Slater determinant explicitly for a two-electron atom, in terms of the radial wave functions and the spherical harmonics in the Schrödinger representation and the spin state functions in the Heisenberg representation of a hydrogenic atom.

7.3. What are the total orbital and spin angular momentum quantum numbers of the ground-state helium and lithium atoms?

7.4. Give the ground state configurations of carbon and silicon. What is the degeneracy of each of these configurations?

7.5. Write the ground state configurations of Ga and As.

8 Interaction of atoms with electromagnetic radiation

The study of interaction of electromagnetic radiation with atoms played a crucial role in the development of quantum mechanics and forms the basis of such important fields of study as spectroscopy, quantum optics, electro-optics, and many important modern devices, such as photo-detectors and lasers. Because the electromagnetic fields acting on the atom are time-varying parameters, the corresponding Schrödinger equation is a partial differential equation with time-varying coefficients. As such, it can only be solved by approximate methods, in general. The standard technique of time-dependent perturbation theory for solving such problems is introduced in this chapter. The absorption and emission processes due to electric dipole interaction of atoms with electromagnetic radiation and the related "transition probabilities" and "selection rules" can be understood on the basis of the first order perturbation theory. An important application of the theory is the process of Light Amplification by Stimulated Emission of Radiation (LASER).

8.1 Schrödinger's equation for electric dipole interaction of atoms with electromagnetic radiation

For the present discussion, we consider the electric dipole interaction of an atom with a monochromatic transverse electromagnetic wave with a wavelength λ, long compared with the spatial extent of the atom. It is assumed that the electric field of the wave is a known applied field of the form:

$$\vec{E}(t) = \vec{E}e^{-i\omega t} + \vec{E}^* e^{i\omega t} \tag{8.1}$$

and is not modified by its interaction with the atom. Thus, the Hamiltonian of the atom in the field is of the form:

$$\hat{H} = \hat{H}_0 + \hat{V}_1, \tag{8.2}$$

where V_1 is the electric dipole interaction energy between the atom and the field and is equal to:

$$\hat{V}_1 = -\hat{\vec{P}} \cdot \vec{E}(t). \tag{8.3}$$

$\hat{P}$ is the operator corresponding to the electric dipole operator of the atom and $\hat{H}_0$ is the Hamiltonian of the atom in the absence of the externally applied field. For a single-electron atom at $\vec{r} = 0$, and assuming that the electromagnetic wave is propagating in the x direction and polarized in the z direction, the electric dipole interaction term in the Schrödinger representation is:

$$\hat{V}_1(z, t) = ez\, E_z(t) = ez\, \tilde{E}_z e^{-i\omega t} + ez\tilde{E}_z^* e^{+i\omega t}. \tag{8.4}$$

For the single-electron hydrogenic atom or ion, the corresponding time-dependent Schrödinger equation is of the form:

$$i\hbar \frac{\partial}{\partial t} \Psi(\vec{r}, t) = \hat{H}(\vec{r}, t)\Psi(\vec{r}, t) = [\hat{H}_0(\vec{r}) + \hat{V}_1(z, t)]\Psi(\vec{r}, t)$$

$$= \left\{ [-\frac{\hbar^2}{2m}\nabla^2 + \hat{V}(r)] + \hat{V}_1(z, t) \right\} \Psi(\vec{r}, t). \tag{8.5}$$

Because of the z and t dependences in the $\hat{V}_1(z, t)$ factor, the method of separation of variables cannot be used and Eq. (8.5) becomes impossibly difficult to solve. Fortunately, if the intensity is not too high and the applied electric field amplitude is small compared to the Coulomb field experienced by the electron in the atom, the effect of $\hat{V}_1$ on the wave function can be considered a small perturbation in comparison with that of $\hat{H}_0$. Thus, the standard time-dependent perturbation theory can be used to find an approximate solution of Eq. (8.5).

8.2   Time-dependent perturbation theory

The time-dependent perturbation technique for solving the time-dependent Schrödinger equation is a powerful general approximation technique. In general, two requirements must be met for any approximate solution to be useful: (1) *The error in the neglected remainder must be demonstrably small*, and (2) *there must be a systemic way to improve the accuracy of the approximate result*. The following procedure leads to such a solution.

If the effect of the perturbation term $\hat{V}_1(z, t)$ is small compared to that of the unperturbed Hamiltonian $\hat{H}_0$, it is assumed the solution can be expanded in a power series of successive orders of "smallness," for which an artifice "ε" is introduced:

$$\Psi = \Psi^{(0)} + \varepsilon\Psi^{(1)} + \varepsilon^2\Psi^{(2)} + \varepsilon^3\Psi^{(3)} + \dots + \varepsilon^n\Psi^{(n)}. \tag{8.6}$$

Consistent with such an expansion, the effects of $\hat{V}_1(z, t)$ on the eigen values and eigen functions of the Hamiltonian $\hat{H}$ are considered an order of ε smaller than those of $\hat{H}_0$ and are identified as such by multiplying it by ε, which can eventually be set to 1:

$$i\hbar \frac{\partial}{\partial t} \Psi = \hat{H}\Psi = [\hat{H}_0 + \varepsilon\hat{V}_1]\Psi. \tag{8.5a}$$

Substituting (8.6) into (8.5a) and equating terms of the same order term-by-term, one obtains a hierarchy of equations of successive orders of ε:

$$\varepsilon^0: \qquad i\hbar \frac{\partial}{\partial t}\Psi^{(0)} - \hat{H}_0\Psi^{(0)} = 0, \tag{8.7a}$$

$$\varepsilon^1: \qquad i\hbar \frac{\partial}{\partial t}\Psi^{(1)} - \hat{H}_0\Psi^{(1)} = \hat{V}_1\Psi^{(0)}, \tag{8.7b}$$

$$\varepsilon^2: \qquad i\hbar \frac{\partial}{\partial t}\Psi^{(2)} - \hat{H}_0\Psi^{(2)} = \hat{V}_1\Psi^{(1)}, \tag{8.7c}$$

$$\vdots$$

$$\varepsilon^n: \qquad i\hbar \frac{\partial}{\partial t}\Psi^{(n)} - \hat{H}_0\Psi^{(n)} = \hat{V}_1\Psi^{(n-1)}. \tag{8.7d}$$

These equations can be solved order-by-order. It is important to note that the basic partial differential equations to be solved for every order are always the same; only the driving term on the right, which depends on the solution of the previous order, changes. Therefore, once the zeroth order problem is solved, one can, in principle, solve the nth order equation and find the solution to Eq. (8.5a) to any order of accuracy systematically. For example, the 0th order equation (8.7a) is the unperturbed time-dependent Schrödinger equation. Once it is solved, the driving term $\hat{V}_1\Psi^{(0)}$ of the first order equation (8.7b) is known. Solving (8.7b) leads to the driving term, $\hat{V}_1\Psi^{(1)}$, of the 2nd order equation (8.7c), and so on. In the final solution, the artifice ε can be set to 1 and the systematic approximate solution is:

$$\Psi = \lim_{\varepsilon \to 1}\left\{ \Psi^{(0)} + \varepsilon\Psi^{(1)} + \varepsilon^2\Psi^{(2)} + \varepsilon^3\Psi^{(3)} + \cdots + \varepsilon^n\Psi^{(n)} \right\}. \tag{8.8}$$

Terminating the series at the nth term gives an nth-order solution, whose error is of the $(n+1)$th order. Furthermore, solutions of equation (8.7d) of successively higher orders following this procedure systematically will, in principle, improve the accuracy of the solution of the time-dependent Schrödinger equation. Thus, both criteria of a legitimate approximation procedure are formally met. In practice, however, such an approximation procedure should be applied with caution beyond the lowest few orders and in the limit of large t.

The first order solution according to the above perturbation procedure leads to the famous "Fermi golden rule." An important example of the application of such a perturbation technique is in the problem of resonant emission and absorption of electromagnetic radiation by atomic systems, which is discussed in detail in the following section.

8.3 Transition probabilities

We return now to the problem of interaction of electromagnetic radiation with a hydrogenic atom, as formulated in Section 8.1. Applying the time-dependent perturbation theory to Eq. (8.5) gives the zeroth and first order equations in the Schrödinger representation:

$$\varepsilon^0: \qquad [i\hbar\frac{\partial}{\partial t} + \frac{\hbar^2}{2m}\nabla^2 - V(r)]\Psi^{(0)}(\vec{r}, t) = 0, \qquad (8.9a)$$

$$\varepsilon^1: \qquad [i\hbar\frac{\partial}{\partial t} + \frac{\hbar^2}{2m}\nabla^2 - V(r)]\Psi^{(1)} = V_1\Psi^{(0)}. \qquad (8.9b)$$

Let us assume that the initial condition is that, at $t = 0$, the system is in the state $\Psi_{E_i}(\vec{r})$, or:

$$\Psi^{(0)}(\vec{r}, t = 0) \equiv \Psi_i(\vec{r}, t = 0) = \Psi_{E_i}(\vec{r}), \qquad (8.10)$$

assuming that the relevant time-independent Schrödinger equation:

$$\left[\frac{\hbar^2}{2m}\nabla^2 - V(r)\right]\Psi_{E_i}(\vec{r}) = -E_i\,\Psi_{E_i}(\vec{r})$$

is solved. From (2.21), the solution of Eq. (8.9a) is then:

$$\Psi_i^{(0)}(\vec{r}, t) = \Psi_{E_i}(\vec{r})e^{-\frac{i}{\hbar}E_i t}. \qquad (8.11)$$

Substituting (8.11) into Eq. (8.9b) gives:

$$\left[i\hbar\frac{\partial}{\partial t} + \frac{\hbar^2}{2m}\nabla^2 - V(r)\right]\Psi^{(1)}(\vec{r}, t) = V_1\Psi_{E_i}(\vec{r})e^{-\frac{i}{\hbar}E_i t}. \qquad (8.12)$$

For (8.12), because the differential operator involves terms of separate independent variables $\vec{r}$ and t, the method of separation of variables applies, and the general solution is of the form:

$$\Psi^{(1)}(\vec{r}, t) = \sum_j C_{ij}^{(1)}(t)\Psi_{E_j}(\vec{r})e^{-\frac{i}{\hbar}E_j t}. \qquad (8.13)$$

Substituting (8.13) into (8.12) followed by multiplying the resultant equation by $\Psi_{E_j}^*(\vec{r})$ from the left and integrating over the space coordinates show that the expansion coefficient $C_{ij}^{(1)}(t)$ satisfies the equation:

$$i\hbar\frac{\partial}{\partial t}C_{ij}^{(1)}(t) = \int \Psi_j^*(\vec{r})\,V_1\Psi_{E_i}(\vec{r})e^{\frac{i}{\hbar}(E_j - E_i)t}d\vec{r}, \qquad (8.14)$$

making use of the orthonormality condition of the eigen functions. Equation (8.14) is a simple ordinary differential equation. Its solution is:

$$C_{ij}^{(1)}(t) = -\frac{i}{\hbar} \int\limits_0^t \left[\int \Psi_j^*(\vec{r}) \; V_1(\vec{r}, t') \Psi_{E_i}(\vec{r}) d\vec{r} \right] e^{\frac{i}{\hbar}(E_j - E_i)t'} dt', \tag{8.15}$$

which satisfies the initial condition $C_{ij}^{(1)}(t = 0) = 0$ from (8.10). For the particular perturbation term of the harmonic type given in (8.4), $C_{ij}^{(1)}(t)$ is explicitly:

$$C_{ij}^{(1)}(t) = \frac{ez_{ij}\tilde{E}_z(1 - e^{i(\omega_{ji}-\omega)t})}{\hbar(\omega_{ji} - \omega)} + \frac{ez_{ij}\tilde{E}_z(1 - e^{i(\omega_{ji}+\omega)t})}{\hbar(\omega_{ji} + \omega)}$$

$$\cong \frac{ez_{ij}\tilde{E}_z(1 - e^{i(\omega_{ji}-\omega)t})}{\hbar(\omega_{ji} - \omega)}, \tag{8.16}$$

in the "near-resonance" case where $\omega_{ji} + \omega \gg |\omega_{ji} - \omega| \approx 0$, assuming $E_j > E_i$, corresponding to the absorption process. Thus, to the first order, the formal solution of Eq. (8.5) that satisfies the initial condition (8.10) is:

$$\Psi(\vec{r}, t) = \Psi_{E_i}(\vec{r})e^{-\frac{i}{\hbar}E_i t} + \sum_{j \neq i} C_{ij}^{(1)}(t) \; \Psi_{E_j}(\vec{r})e^{-\frac{i}{\hbar}E_j t} + O(\varepsilon^2)$$

$$\cong \Psi_{E_i}(\vec{r})e^{-\frac{i}{\hbar}E_i t} + \sum_{j \neq i} \left[\frac{ez_{ij}\tilde{E}_z(1 - e^{i(\omega_{ji}-\omega)t})}{\hbar(\omega_{ji} - \omega)} \right] \Psi_{E_j}(\vec{r})e^{-\frac{i}{\hbar}E_j t} + O(\varepsilon^2), \tag{8.17}$$

where

$$z_{ij} = \left[\int \Psi_{E_j}^*(\vec{r}) \; z \Psi_{E_i}(\vec{r}) d\vec{r} \right] \tag{8.18}$$

is the z-component of the "induced electric dipole moment," or the "transition moment," between the eigen states Ψ_{E_i} and Ψ_{E_j}. From parity considerations, $z_{ii} \equiv 0$ (see also Section 8.4); the ith term is, therefore, excluded from the sum in (8.17). The physical interpretation of this very important result, (8.17), is somewhat tricky.

Equation (8.17) shows that there is a certain probability that an applied electric field can induce a transition of the atom from the initial state Ψ_{E_i} to the state Ψ_{E_j}. According to the interpretation of the wave function, the probability of finding the atom in the state Ψ_{E_j} at time t is:

$$\left| C_{ij}^{(1)}(t) \right|^2 = \frac{e^2}{\hbar^2} |z_{ij}|^2 |\tilde{E}_z|^2 \; \frac{2 - 2\cos(\omega_{ji} - \omega)t}{(\omega_{ji} - \omega)^2}, \tag{8.19}$$

where $j \neq i$. A "transition probability," corresponding to the probability per unit time an atom initially in the state Ψ_{E_i} is induced to make a transition to the Ψ_{E_j} state, can be defined:

$$W_{ij} \equiv \frac{\partial \left| C_{ij}^{(1)}(t) \right|^2}{\partial t} = \frac{e^2}{\hbar^2} \left| z_{ij} \right|^2 \left| \tilde{E}_z \right|^2 \left[\frac{2 \sin(\omega_{ji} - \omega)t}{(\omega_{ji} - \omega)} \right]. \tag{8.20}$$

The last factor is proportional to the Dirac delta-function in the limit of $t \to \infty$:

$$\lim_{t \to \infty} \frac{\sin(\omega_{ji} - \omega)t}{(\omega_{ji} - \omega)} = \pi \, \delta(\omega_{ji} - \omega),$$

because, near where $(\omega_{ji} - \omega) \sim 0$, it increases as t approaches ∞, and it decreases rapidly to exactly 0 at $(\omega_{ji} - \omega) = \pm\pi/t$, and to essentially zero (relative to the peak) beyond. The area under the peak between $(\omega_{ji} - \omega) = \pm\pi/t$ is approximately equal to π. Thus, the probability of transition from the state Ψ_{E_i} to the state Ψ_{E_j} per unit time induced by the monochromatic incident wave on the atoms is:

$$W_{ij} = \frac{2\pi e^2}{\hbar^2} \left| z_{ij} \right|^2 \left| \tilde{E}_z \right|^2 \delta(\omega_{ji} - \omega), \tag{8.21a}$$

which shows the important resonance condition that the frequency ω of the incidence wave must be equal to the transition frequency ω_{ij} of the atom, and that transition probability is linearly proportional to the intensity of the incident wave. Equation (8.21a) is a form of the "Fermi golden rule." Since the energy of the photon is $\hbar\omega$, the resonance condition shows that energy is conserved in the single-photon absorption process. The atom can only absorb one photon of energy $\hbar\omega = \hbar\omega_{ji}$ at a time while being promoted from the Ψ_{E_i} to the Ψ_{E_j} state. Similarly, if $E_j < E_i$, the corresponding process corresponds to the spontaneous emission of a single photon of energy $\hbar\omega$ while the atom drops from the state Ψ_{E_j} to the state Ψ_{E_i} with the transition probability:

$$W_{ij} = \frac{2\pi e^2}{\hbar^2} \left| z_{ij} \right|^2 \left| \tilde{E}_z \right|^2 \delta(\omega_{ij} - \omega). \tag{8.21b}$$

Equations (8.21a) and (8.21b) are derived for radiative transitions between sharply defined energy levels induced by monochromatic waves. In practical situations, the finite widths of the radiation spectrum and the transition frequency range must be taken into account.

If the transition frequency is not sharply defined, either because the lifetimes of the initial and final states are finite or because of the slight variation in the local environment of the atoms in a macroscopic sample, then radiative transition can take place over a range of frequencies. The corresponding "line shape function" is not a delta-function as in (8.21a) or (8.21b) but some normalized general distribution function $g(\omega_{ij} - \overline{\omega}_{ij})$ centered on $\overline{\omega}_{ij}$, where $\int g(\omega - \overline{\omega}_{ij}) \, d\omega = 1$. If the energy levels of the initial and final states of the radiative induced transition are broadened because of the finite lifetimes of these states only, the mechanism is called "homogeneous broadening" and the corresponding line shape function $g(\omega_{ij} - \overline{\omega}_{ij})$ is "Lorentzian," as will be discussed in detail in Chapter 11. If the energy levels are broadened because of the local environmental variations, it is called "inhomogeneous broadening" and the line shape function $g(\omega_{ij} - \overline{\omega}_{ij})$ tends to be "Gaussian."

In the case of spontaneous emission, or fluorescence, from the upper level E_i, the transition probability must be integrated over the transition frequency:

$$W_{ij} = \frac{2\pi \, e^2}{\hbar^2} |z_{ij}|^2 |\tilde{E}_z|^2 \int g(\omega_{ij} - \overline{\omega}_{ij}) \, \delta(\omega - \omega_{ij}) \, d\omega_{ij}$$

$$= \frac{2\pi \, e^2}{\hbar^2} |z_{ij}|^2 |\tilde{E}_z|^2 \, g(\omega - \overline{\omega}_{ij}), \tag{8.21c}$$

and the fluorescence line shape function $g_f(\omega - \omega_{ij})$ is of the form $g(\omega - \overline{\omega}_{ij})$.

In the case of resonance absorption from the lower energy level E_i and the incident radiation being not a monochromatic wave but having a normalized spectrum of the form $\rho(\omega)$, the transition probability in (8.21a) must be integrated over both the distribution of the transition frequency and incident radiation spectrum:

$$W_{ij} = \frac{2\pi \, e^2}{\hbar^2} |z_{ij}|^2 |\tilde{E}_z|^2 \int \int g_f(\omega_{ij} - \overline{\omega}_{ij}) \, \rho(\omega) \, \delta(\omega - \omega_{ij}) d\omega_{ij} \, d\omega$$

$$= \frac{2\pi \, e^2}{\hbar^2} |z_{ij}|^2 |\tilde{E}_z|^2 \int g_f(\omega_{ij} - \overline{\omega}_{ij}) \, \rho(\omega_{ij}) \, d\omega_{ij}. \tag{8.21d}$$

Thus, if the fluorescence line width is much narrower than the spectral width of the incident radiation, the transition probability for absorption, (8.21d), reduces to:

$$W_{ij} \cong \frac{2\pi e^2}{\hbar^2} |z_{ij}|^2 |\tilde{E}_z|^2 \, \rho(\overline{\omega}_{ij}). \tag{8.21e}$$

If the spectral width of the incident radiation is much narrower than the fluorescence line width, the transition probability (8.21d) becomes:

$$W_{ij} \cong \frac{2\pi \, e^2}{\hbar^2} |z_{ij}|^2 |\tilde{E}_z|^2 \, g_f(\omega_0 - \overline{\omega}_{ij}), \tag{8.21f}$$

where ω_0 is the center-frequency of the incident radiation.

Equations (8.21b)–(8.21f) are the Fermi golden rule for radiative transitions.

8.4   Selection rules and the spectra of hydrogen and hydrogen-like ions

Equations (8.21a) and (8.21b) show that the transition probability for the absorption or emission process between the initial state Ψ_{E_i} and the final state Ψ_{E_j} depends on the magnitude of the induced matrix element defined in (8.18):

$$z_{ij} = \int \Psi^*_{E_j}(\vec{r}) \, z \, \Psi_{E_i}(\vec{r}) \mathrm{d}\vec{r}. \tag{8.18}$$

Thus, whether a particular transition is allowed or not depends on the spatial symmetry of the wave functions of the initial and final states in the spatial integral defining the induced matrix element z_{ij}.

For example, for the case of linearly polarized wave, induced transition can take place only between states of opposite parity, as we will now show. Since the integration in (8.18) is to be carried out over all space, the integral should be invariant under inversion of the coordinate axes; thus,

$$\begin{aligned} z_{ij} &= \int \Psi^*_{E_j}(\vec{r}) \, z \, \Psi_{E_i}(\vec{r}) \mathrm{d}\vec{r} \\ &= \int \Psi^*_{E_j}(-\vec{r})(-z) \, \Psi_{E_i}(-\vec{r}) \mathrm{d}\vec{r}. \end{aligned} \tag{8.22}$$

Making use of the concept of parity operator defined previously in Eq. (4.31), (8.22) becomes:

$$\begin{aligned} z_{ij} &= \int \Psi^*_{E_j}(-\vec{r})(-z) \, \Psi_{E_i}(-\vec{r}) \mathrm{d}\vec{r} \\ &= \int [\hat{P}\Psi_{E_j}(\vec{r})]^* \, (-z) \, [\hat{P}\Psi_{E_i}(\vec{r})] \, \mathrm{d}\vec{r}; \end{aligned}$$

thus, the product of the eigen values of the parity operator $\hat{P}$ corresponding to the eigen states Ψ_{E_i} and Ψ_{E_j}, respectively, must be equal to -1, and *the states Ψ_{E_i} and Ψ_{E_j} must be of opposite parity*. Similar considerations apply to the x and y components of the transition matrix element. This is one of the "selection rules" for the emission and absorption processes.

There are other rules depending on other symmetry properties such as the angular symmetry properties of the wave functions involved. For example, suppose the angular parts of the initial and final wave functions in (8.18) are $Y_{\ell m_\ell}(\theta, \phi)$ and $Y_{\ell' m'_\ell}(\theta, \phi)$, respectively. Analogous to the parity consideration, integration of the coordinate variable ϕ leads to the selection rules on the azimuthal quantum numbers m_ℓ and m'_ℓ:

$$\Delta m_\ell \equiv m_\ell - m'_\ell = 0 \text{ for waves linearly polarized in the } z\text{-direction}; \tag{8.23a}$$

and

$$\Delta m_\ell \equiv m_\ell - m'_\ell = \pm 1 \text{ for right and left circularly polarized waves.} \tag{8.23b}$$

Table 8.1. *Approximate measured wavelengths in air (in nm except as otherwise indicated) of some of the discrete lines in the spectrum of hydrogen. (See, for example, Herzberg (1944). More precise values can be found from the data in the US National Institute of Standards and Technology Handbooks on Atomic Energy Levels.)*

n =	1	2	3	4	5
n′ = 2	121.6				
3	102.6	656.3			
4	97.3	486.1	1875.1		
5	95.0	434.0	1281.8	4.06 μm	
6	93.8	414.1	1093.8	2.63 μm	7.40 μm
7		397.0	1005.0		
8		388.9	954.6		
9		383.5			
10		379.8			

For (8.23b), the axis of quantization of the atomic wave function is perpendicular to the plane of polarization of the incident wave.

These selection rules reflect the conservation of angular momentum in the emission and absorption process, since the angular momentum of the circularly polarized photons is $\pm\hbar$, and the linearly polarized photon is an equal mixture of the photon states with $\pm\hbar$ angular momentum relative to the axis of quantization of the atomic wave functions. Similar considerations in θ involving associated Legendre functions lead to the selection rule on the orbital quantum numbers ℓ and ℓ' for dipole induced transitions:

$$\Delta\ell \equiv \ell - \ell' = \pm 1. \tag{8.23c}$$

Thus, the selection rules depend on the nature of the quantum states involved in the transition and the state of polarization of the radiation. Such rules and the resonance condition are the key considerations that determine the general features of the emission and absorption spectra of all atoms, molecules, and solids.

Consider, for example, the discrete absorption spectra of hydrogen and hydrogen-like ions (Z protons in the nucleus and one electron) initially in the ground 1s state. The selection rule (8.23c) for the orbital quantum numbers shows that from this ground state, the atom can absorb a photon and be excited into one of the quantized np levels, where $n = 2, 3, 4 \ldots$ For the hydrogen atom in particular, $Z = 1$, and the corresponding wavelengths of the discrete absorption lines are:

$$\frac{1}{\lambda_{1s,np}} = R_H Z^2 \left(1 - \frac{1}{n'^2}\right), \text{ for } 1 < n' = 2, 3, 4, \ldots, \tag{8.24}$$

where $R_H = \dfrac{me^4}{4\pi c\hbar^3}$, from (6.37), is the Rydberg constant and is numerically equal to

109 737.3 cm^{-1}. The longest wavelength of this series of discrete absorption lines is, therefore, 121.566 nm in the ultraviolet. These absorption lines and the corresponding fluorescence emission lines (np $\rightarrow$ 1s) form the so-called "Lyman series" of the hydrogen spectrum and are tabulated in Table 8.1.

Based on the model of the hydrogen-like ions in general given in this chapter, the wavelengths of the discrete line spectra corresponding to the transitions between other energy eigen states of the hydrogen atom ($Z = 1$) subject to the selection rule (8.23c) satisfy the Rydberg formula:

$$\frac{1}{\lambda_{n\ell,n'\ell\pm1}} = R_H\left(\frac{1}{n^2} - \frac{1}{n'^2}\right), \text{ where } n = 1, 2, 3, \ldots \text{ and } n' > n, \tag{8.24a}$$

including (8.24) for the Lyman series ($n=1$). The series with $n=2, 3, 4, 5, \ldots$ correspond to the Balmer ($n=2$), Ritz-Paschen ($n=3$), Bracket ($n=4$), Pfund ($n=5$), ... series, respectively. Examples of the experimentally observed wavelengths in air of some of these lines are also tabulated in Table 8.1

8.5   The emission and absorption processes

A simple picture of the emission and absorption processes can be given on the basis of the formal mathematical solutions developed in the previous section. For definiteness, let us consider the specific example of the hydrogen atom. Suppose the atom is initially in the 1s level. Since the electric dipole interaction term V_1 in the Hamiltonian does not involve the spin of the atom, we can neglect the spin quantum numbers in labeling the wave functions; thus, the initial state is the $|100\rangle$ state, and the final states are the $|n\ell m_\ell\rangle$ states, of the hydrogen atom.

The probability distribution function of the electron in the $|100\rangle$ state of the hydrogen atom is shown schematically in Figure 6.5(a). It is spherically symmetrically centered on the positively charged nucleus and has even parity. Therefore, the atom in the 1s state has no electric dipole moment and does not interact with any applied electric field if it remains in the ground state. In fact, the probability distribution of the electron in any unperturbed energy eigen state of the atom is always invariant under coordinate inversion $\vec{r} \rightarrow -\vec{r}$ because the potential term in the Hamiltonian is invariant under the same inversion of the coordinate system. Thus, the atom in an unperturbed energy eigen state cannot have any electric dipole moment. For the atom to have an electric dipole moment, the atomic wave function must be in a superposition state of mixed parity. Consider, for example, an applied electric field polarized in the z direction. The selection rules (8.23a) and (8.23c) dictate that, for a 1s initial state, the state function in the presence of the incident field in the single-photon absorption process must be, for example, a superposition of the 1s or $|100\rangle$ state and the $|210\rangle$ or $2p_z$ state (for a linearly polarized wave), which has odd parity:

$$|E_f\rangle = |100\rangle\, e^{-\frac{i}{\hbar}E_1 t} + C_{12}^{(1)}|210\rangle, e^{-\frac{i}{\hbar}E_2 t}. \tag{8.25}$$

The expectation value of the induced electric dipole moment, P_z, of the atom in this mixed state is finite:

$$P_z = \langle E_f|(-ez)|E_f\rangle = C_{12}^{(1)}\langle 100|(-ez)|210\rangle e^{-i\omega_{21}t} + \text{complex conjugate},$$

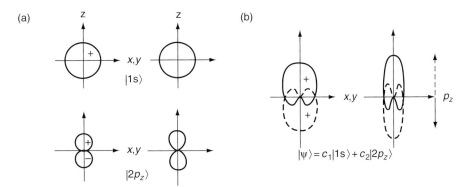

Figure 8.1. *Schematics showing the wave functions (left, the $+$ and $-$ signs refer to the numerical values of the wave functions) and the corresponding probability distribution functions (right) of (a) the energy eigen states, and (b) the mixed state (solid curves: $t = 0$, $2\pi/\omega_{21}$, $4\pi/\omega_{21}, \ldots$; dashed curves: $t = \pi/\omega_{21}$, $3\pi/\omega_{21}, \ldots$) of the wave functions shown in (a). As the charge distribution oscillates up and down, so will the induced dipole moment P_z oscillate up and down at the frequency ω_{21}.*

where $C_{12}^{(1)}$ is given by (8.16). The amplitude of this induced dipole moment is the largest at resonance $\omega = \omega_{21}$. In this limit, (8.17) shows that it increases with t. It also shows that, in this limit, the dipole moment oscillates at the angular frequency $\omega_{21} = \omega$, lags the applied electric field in phase by $\pi/2$, and is proportional to the amplitude of the incident field and the transition moment $\langle 100|z|210 \rangle$. This is the physical basis of the single-photon absorption process.

This induced absorption process can also be understood qualitatively, as shown in Figure 8.1. The wave functions and the corresponding charge distribution functions of the 1s and 2p$_z$ states are shown schematically in Figure 8.1(a). In both cases, the charge distribution functions are symmetrically located relative to the positively charged nucleus. The wave function and charge distribution function corresponding to (8.25) are shown schematically in Figure 8.1(b). It is clear that, because the two components have opposite parity, the distribution function corresponding to the sum of the two wave functions is skewed in the z direction relative to the nucleus and, therefore, the atom in the mixed state has an induced dipole moment. At resonance, $\omega = \omega_{21}$, when $t = \pi/\omega$, the phase of the 2p$_z$ state changes by π relative to that of the 1s wave function, the resultant charge distribution function now becomes skewed in the opposite direction and the direction of the induced dipole reverses. This is analogous to the wave packet oscillation phenomenon discussed in Section 5.3 It repeats every cycle, leading to a larger and larger oscillating dipole at the frequency ω of the applied field and a bigger and bigger 2p$_z$ component in the mixed state. This is the basic quantum mechanic picture of the resonant absorption process of the atom.

Suppose the energy of the initial state is above that of the final state, or $E_i > E_j$. For example, if the hydrogen atom is initially in the 2p$_z$ state and the final state is the 1s state, the term with $(\omega_{ji} - \omega)$ in the denominator in (8.16) should be replaced by the resonant term with $(\omega_{ji} + \omega)$ in the denominator, leading to an emission process. The resulting induced dipole moment will lead the applied field in phase by $\pi/2$. If the existing field is the externally applied field of an incident wave, the field emitted by the oscillating dipole will be in phase with and add to the incident field. The corresponding emission process is the "stimulated emission process," as will be discussed in more detail in the following section. If the only field present is the vacuum fluctuation field (see the discussion following (5.21)), the emission process is the "spontaneous emission process" which is responsible for the "fluorescence" or "luminescence" spectra of the atom in the excited non-equilibrium states. The corresponding emitted field will have random phases reflecting those of the vacuum fluctuation fields.

8.6   Light Amplification by Stimulated Emission of Radiation (LASER) and the Einstein A- and B-coefficients

One of the most important practical consequences of the quantum theory of the process of light emission by atomic systems is the development of the ubiquitous laser. (See, for example, Siegman (1986) or Yariv (1989).) As shown in the previous section, when the atom is initially in an excited state, it can be stimulated to emit a photon of the same frequency and phase as that of the incident photon at resonance. Furthermore, if the phase of the incident wave is well defined, as in a classical wave of the form (8.1), the emitted wave will be in phase with and add to the incident wave coherently. This is the stimulated emission process.

Consider, for example, an incident monochromatic plane wave, polarized in the x direction, of the form:

$$\vec{E}(z, t) = \left[\tilde{E}_x e^{i(kz - \omega t)} + \tilde{E}_x^* e^{-i(kz - \omega t)} \right] E_x \tag{8.26}$$

propagating in the z direction in a macroscopic medium of "two-level" atoms with energies E_2 and E_1, where $(E_2 - E_1)/\hbar = \omega_{21} \sim \omega$. Real atoms have, of course, many energy levels. In the resonant emission or absorption process, only the initial and final states are directly affected by the interaction process. We can, therefore, focus only on these two relevant energy levels of the atom. Let us assume that, in the absence of the incident wave, the medium is in an equilibrium state and that there are N_2 atoms per unit volume in the upper level and N_1 atoms per unit volume in the lower level. In the presence of the incident wave, the atoms in the lower level will absorb photons and be excited into the upper level and the atoms in the upper level will emit photons and drop down to the lower level. If N_1 is greater than N_2, there will be net absorption of photons. If N_2 is greater than N_1, there will be net emission of photons. The rates of change of the "populations" of the atoms in the upper and lower levels due to such photon-induced transitions are, therefore, respectively:

$$\frac{dN_2}{dt} = -N_2 W_{21} + N_1 W_{12},$$
$$\frac{dN_1}{dt} = N_2 W_{21} - N_1 W_{12}, \tag{8.27}$$

where $W_{12} = W_{21}$ is defined in (8.21a) or (8.21b). The net rate of change of the "population inversion" $(N_2 - N_1)$ is, therefore:

$$\frac{d}{dt}(N_2 - N_1) = -2(N_2 - N_1)W_{21} \qquad (8.28)$$

and the corresponding change in the photon numbers N_{ph} per unit volume is:

$$\frac{dN_{ph}}{dt} = (N_2 - N_1)W_{21}.$$

If the incident wave is not a monochromatic wave and the spectral density of the incident radiation is $\rho(\nu)$, where $\int \rho(\nu)d\nu = 1$, the rate of change of the volume density of the light wave energy $\varepsilon \equiv |\tilde{E}_x|^2/2\pi$ in the medium is, from (8.21d):

$$\frac{d}{dt}\varepsilon = (N_2 - N_1)\frac{4\pi^2 e^2 \nu}{\hbar}|x_{12}|^2 \rho(\nu_{21})\, \varepsilon. \qquad (8.29)$$

In a real medium, there may be other processes taking place in the medium, so that the transition probability is spectrally broadened from the delta-function dependence shown in (8.21a & b) into a fluorescence line shape function $g_f(\nu)$, where $\int g_f(\nu)\, d\nu = 1$. In that case, if the line width of the incident radiation is much narrower than the fluorescence line width and can be considered a monochromatic wave of frequency ν, the spectral line shape function $\rho(\nu_{21})$ of the radiation in (8.29) should be replaced by the fluorescence line shape function $g_f(\nu)$ evaluated at the frequency ν, as shown in (8.21f):

$$\frac{d}{dt}\varepsilon = (N_2 - N_1)\frac{4\pi^2 e^2 \nu}{\hbar}|x_{12}|^2 g_f(\nu)\, \varepsilon, \qquad (8.30)$$

Converting this to the spatial rate of change of the intensity of the wave gives the spatial gain coefficient, g, in the medium:

$$\frac{\partial I}{\partial z} = gI = (N_2 - N_1)\,\sigma_{st}\, I, \qquad (8.31)$$

where $\sigma_{st} = \frac{4\pi^2 e^2 \nu}{\hbar c}|x_{12}|^2 g_f(\nu)$ and c is the velocity of the wave in the medium. σ_{st} is known as "the stimulated emission cross section." This equation shows the important result that the electromagnetic wave will be "amplified" if there is population inversion in the medium, i.e. $N_2 > N_1$. If such an amplifying medium is enclosed in a suitable electromagnetic cavity in which most of the emitted radiation can be reflected back into the medium along the same path again and again for repeated amplification, even a small amount of initially present spontaneous emission can grow into a powerful coherent beam of stimulated emission at the frequency $\omega \approx \omega_{21}$. This is the basis of the "laser" with its numerous practical applications.

Einstein A- and B-coefficients

Instead of the cross section, the stimulated emission process is often characterized by the well known "Einstein B-coefficient," which is by definition:

$$B_{12} \equiv \frac{2\pi e^2}{\hbar^2}|x_{12}|^2 = \sigma_{st}c/h\nu\, g_f(\nu). \tag{8.32}$$

There is also an "Einstein A-coefficient," which characterizes the "spontaneous emission process." It has to do with the fact that, if an atom is in an excited state, it must eventually drop down to an available lower energy state. Take the two-level atom with allowed radiative dipole transition between the two levels 2 and 1 as an example again. The rate of change of the population in the upper level must have a decay term even in the absence of any applied radiation:

$$\frac{d}{dt}N_2 = -\frac{N_2}{\tau_2}, \tag{8.33a}$$

and there is a corresponding rate of increase term for the population of the lower state:

$$\frac{d}{dt}N_1 = +\frac{N_2}{\tau_2}. \tag{8.33b}$$

This "relaxation time" τ_2 gives the "radiative life time" of level 2 (this population relaxation time τ_2 for 'level 2' is not to be confused with the "atomic coherence time T_2" to be introduced in Chapter 11), if there is no other lower energy level the atom can decay to. The corresponding "relaxation rate" τ_2^{-1} is by definition the "Einstein A-coefficient" A_{21}. If there are n levels of lower energies to which the atom in a higher energy level i can decay, then the total radiative decay rate out of the ith level is:

$$A_i = \sum_{i=1}^{n} A_{ij}, \tag{8.34}$$

and the radiative life time of the ith level τ_i is equal to A_i^{-1}.

The Einstein A- coefficient is directly related to the Einstein B-coefficient for the same transition. The relationship between the two can be found simply by considering the situation where the two-level atom is in thermal equilibrium with the black-body radiation $\rho_b(\nu,\,_{21})$, as discussed in Section 5.4. The corresponding rate equation for the population in level 2 in the presence of the black-body radiation is, from (5.55), (8.28), (8.32), and (8.33a & b):

$$\frac{d}{dt}(N_2 - N_1) = -2A_{21}N_2 - 2B_{21}(N_2 - N_1)\rho_b(\nu_{21})$$

$$= -2A_{21}N_2 - 2B_{21}(N_2 - N_1)\left[\frac{8\pi\, h\nu_{21}^3}{c^3}\cdot\frac{1}{e^{h\nu_{21}/k_BT} - 1}\right]. \tag{8.35}$$

In thermal equilibrium at the temperature T, we know that $\frac{d}{dt}(N_2 - N_1) = 0$ and the ratio of the populations in the upper and lower energy levels is determined by the Boltzmann factor: $e^{-\frac{h\nu_{21}}{k_B T}}$. Thus, from (8.35), the ratio of the Einstein coefficients must be:

$$\frac{A_{21}}{B_{21}} = \frac{8\pi h \nu_{21}^3}{c^3}. \tag{8.36}$$

This is a very important result which shows that the B-coefficient, or the induced dipole matrix element $|x_{21}|^2$, can be determined from the A-coefficient, which can in turn be determined experimentally from the measured corresponding radiative life time of the atoms in level 2. The Einstein B-coefficient determines the stimulated emission cross section and, hence, the gain coefficient in the laser.

In this section, in considering the interaction of electromagnetic radiation with a uniform macroscopic medium of N two-level atoms per unit volume, it is assumed that all the atoms are independent of each other and that there are exactly N_1 atoms in the lower level and N_2 atoms in the upper level. The net absorption per unit volume of the medium is, therefore, $(N_1 - N_2)$ times the absorption cross section per atom. It often happens that, in practical situations involving the interaction of optical media with coherent electromagnetic radiation, such as the laser light, it is not possible to know the exact state of the N-particle system. The most that can be known and specified is the probability distribution function P_Ψ over all the possible states $|\Psi\rangle$ the N atoms in the macroscopic medium can be in. Furthermore, the optical properties of the medium under intense coherent light often depend on the collective response of the atoms in the medium. For such problems, the state and the dynamics of the medium are generally analyzed using the density-matrix formulation and the quantum mechanic Boltzmann equation, which will be discussed in detail in Chapter 11

8.7 Problems

8.1 Verify that the result given in Eq. (8.17) indeed satisfies the time-dependent Schrödinger equation (8.5) to the first order of V_1.

8.2 Derive the transition probability analogous to (8.21b) for right- and left-circularly polarized electromagnetic waves.

8.3 Verify the selection rules given in (8.23a–c) for hydrogenic ions.

8.4 Compare the experimentally observed Lyman series discrete spectra for hydrogen given in Table 8.1 with the predictions of the Rydberg formula (8.24).

8.5 Give the expectation value of the z component of the electric dipole moment of the hydrogen atom in the mixed state:

$$\Psi(\vec{r}, t) = \frac{1}{\sqrt{1 + |C_{12}|^2}} [\Psi_{100}(\vec{r}, t) + C_{12}\Psi_{210}(r, \theta, \phi; t)].$$

8.6 An electron in the $n=3$, $\ell=0$, $m=0$ state of hydrogen decays by a sequence of (electric dipole) transitions to the ground state.

(a) What decay routes are open to it? Specify them in the following way: $|300\rangle \rightarrow |n\ell m\rangle \rightarrow |n'\ell'm'\rangle \cdots \rightarrow |100\rangle$.

(b) What are the allowed transitions from the 5d states of hydrogen to lower states?

8.7 Give the stimulated emission cross section (in cm^2) defined in connection with (8.31) for a hypothetical hydrogen laser with linearly polarized emission at 121.56 nm (Lyman-α line). Assume a Lorentzian fluorescence linewidth of 10 Ghz. What is the corresponding spatial gain coefficient (in cm^{-1}) if the total population inversion between the 1s and 2p levels of hydrogen in the gaseous medium is 10^{10} cm^{-3}? (Assume all the degenerate states in the 2p level are equally populated.)

9 Simple molecular orbitals and crystalline structures

With the basic quantum theory of atomic systems developed in the previous chapters, it is now possible to address the question, at least in a qualitative way, of how atoms can be held together to form molecules and crystalline solids. The explanation is based on the time-independent Schrödinger equation, which is solved on the basis of time-independent perturbation theory.

When the atoms are brought together, the electrons and ions in the atoms interact also with the positive charges in the nuclei and the electrons of the neighboring ions. Quantum mechanically, it may be energetically more favorable for the atoms to form molecular complexes than to exist as separate atoms. A simple molecular orbital theory of "covalent bonded" diatomic molecules is introduced. This model can lead to a qualitative understanding of, for example, some simple sp-, sp^2-, or sp^3- bonded organic molecules, and sp^3-bonded tetrahedral complexes that are the basic building blocks of such important IV–IV elemental semiconductors as Si and Ge and various III–V and II–VI compound semiconductors such GaAs, GaP, ZnS, and CdS. The basic geometry of the atomic orbitals of the constituent atoms determines the structures of the tetrahedral complexes, which in turn determine the crystalline structures of the solids. Of particular interest are semiconductors with broad applications in electronics and photonics.

9.1 Time-independent perturbation theory

The key to solving the time-dependent Schrödinger equation is to solve the corresponding time-independent Schrödinger equation. Yet, in the vast majority of cases, the time-independent Schrödinger equation cannot be solved exactly analytically. Time-independent perturbation theory is a powerful rigorous procedure for systematically solving time-independent Schrödinger equations approximately. The general procedure of such a theory is outlined in this section. It will be used to deal with a variety of problems related to molecules and solids in the following sections. The procedures for the non-degenerate states and the degenerate states are different. We will consider these separately in order.

Non-degerate perturbation theory

The more compact Dirac notation is used here. Just like the time-dependent perturbation theory, the basic idea is to separate the Hamiltonian $\hat{H}$ into a large part, the unperturbed Hamiltonian $\hat{H}_0$, and a small part $\hat{H}_1$, which is considered a perturbation and is identified as such with a multiplier ε:

$$\hat{H} = \hat{H}_0 + \varepsilon \hat{H}_1. \tag{9.1}$$

ε will eventually be set equal to 1 in the final results. To find systematically the eigen functions and eigen values approximately, we expand each in a power series of ε:

$$E_i = E_i^{(0)} + \varepsilon E_i^{(1)} + \varepsilon^2 E_i^{(2)} + \dots, \tag{9.2}$$

$$|E_i\rangle = |E_i^{(0)}\rangle + \varepsilon |E_i^{(1)}\rangle + \varepsilon^2 |E_i^{(2)}\rangle + \dots, \tag{9.3}$$

subject to the normalization condition:

$$\langle E_i | E_i \rangle = 1 \tag{9.4}$$

in the corresponding eigen value equation. Substituting (9.1)–(9.3) in the time-independent Schrödinger equation:

$$\hat{H}|E_i\rangle = E_i|E_i\rangle, \tag{9.5}$$

and equating terms of the same order of ε term-by-term leads to a hierarchy of operator equations in successive orders of ε:

$$\varepsilon^0 : \qquad \hat{H}_0|E_i^{(0)}\rangle = E_i^{(0)}|E_i^{(0)}\rangle, \tag{9.5a}$$

$$\varepsilon^1 : \qquad (\hat{H}_0 - E_i^{(0)})|E_i^{(1)}\rangle + (\hat{H}_1 - E_i^{(1)})|E_i^{(0)}\rangle = 0, \tag{9.5b}$$

$$\varepsilon^2 : \qquad (\hat{H}_0 - E_i^{(0)})|E_i^{(2)}\rangle + (\hat{H}_1 - E_i^{(1)})|E_i^{(1)}\rangle - E_i^{(2)}|E_i^{(0)}\rangle = 0, \tag{9.5c}$$

.

.

.

This is the so-called Rayleigh–Schrödinger perturbation procedure (there are other procedures, such as the Brillouin perturbation procedure).

The first step in solving these equations is to solve the unperturbed Schrödinger equation (9.5a). It is assumed that this can been done; otherwise, the procedure will not work. Thus, it is assumed that $E_i^{(0)}$ and $|E_i^{(0)}\rangle$ are known. Multiplying (9.5b) from the left by the bra-vector $\langle E_i^{(0)}|$ gives the first order correction to the eigen value E_i due to the perturbation:

$$E_i^{(1)} = \langle E_i^{(0)}|\hat{H}|E_i^{(0)}\rangle. \tag{9.6a}$$

Multiply (9.5b) from the left by a bra-vector $\langle E_j^{(0)}|$ corresponding to a different eigen state $(j \neq i)$ gives:

$$\langle E_j^{(0)}|E_i^{(1)}\rangle = \frac{\langle E_j^{(0)}|\hat{H}_1|E_i^{(0)}\rangle}{E_i^{(0)} - E_j^{(0)}}, \tag{9.6b}$$

and from (9.4),

$$\langle E_i^{(0)}|E_i^{(1)}\rangle = 0; \tag{9.6c}$$

therefore, the eigen function to the first order is:

$$|E_i\rangle = |E_i^{(0)}\rangle + \sum_{j \neq i} \frac{\langle E_j^{(0)}|\hat{H}_1|E_i^{(0)}\rangle}{E_i^{(0)} - E_j^{(0)}} |E_j^{(0)}\rangle + \dots . \tag{9.6d}$$

A similar procedure will give the higher order corrections of the eigen values and eigen functions. For example, the perturbation solutions (9.2) and (9.3) carried to the second order correction of the eigen value with ε set to 1 is:

$$E_i = E_i^{(0)} + \langle E_i^{(0)}|\hat{H}_1|E_i^{(0)}\rangle + \sum_{j \neq i} \frac{\left|\langle E_j^{(0)}|\hat{H}_1|E_i^{(0)}\rangle\right|^2}{E_i^{(0)} - E_j^{(0)}} + \dots . \tag{9.7}$$

Thus, once the zeroth order equation (9.5a) is solved, it is possible to follow this procedure to obtain rigorously and systematically a perturbative solution to the time-independent Schrödinger equation to an arbitrary order of accuracy, at least in principle.

It should be pointed out, however, that the choice of what should be considered the unperturbed part of the Hamiltonian and what should be considered the perturbation is not unique. One should generally choose $\hat{H}_1$ to be small enough so that only a small number of terms in the perturbation series are needed to give a reasonably accurate answer, and yet it is not so small that the unperturbed part of the problem becomes too difficult to solve.

These results, (9.6d & 9.7), clearly will not be valid for unperturbed eigen states that are degenerate (i.e. different zeroth-order eigen states with the same eigen value, or $E_i^{(0)} = E_j^{(0)}$). When that happens, a special degenerate perturbation theory is needed.

Degenerate perturbation theory

It frequently happens that some unperturbed eigen energy level has degeneracy, because there are other constants of motion in addition to the total energy of the system (see the discussion following Eq. (2.49)). Suppose such a constant of motion is represented by the operator $\hat{B}$ with eigen values β_j, which commutes with the unperturbed Hamiltonian $\hat{H}_0$ of the system. The simultaneous eigen states of the two operators are $|E_i^{(0)}\beta_j\rangle$ so that:

$$\hat{H}_0|E_i^{(0)}\beta_j\rangle = E_i^{(0)}|E_i^{(0)}\beta_j\rangle, \tag{9.8a}$$

and

$$\hat{B}|E_i^{(0)}\beta_j\rangle = \beta_j|E_i^{(0)}\beta_j\rangle. \tag{9.8b}$$

In this case, there is a whole family of degenerate states $|E_i^{(0)}\beta_j\rangle$ that all have the same energy $E_i^{(0)}$. This degeneracy may be partially removed when the perturbation $\hat{H}_1$ is taken into account through the following procedure. First, multiplying (9.5a) by the bra-vector $\langle E_j^{(0)}\beta_k|$, where $E_j^{(0)} \neq E_i^{(0)}$, from the left and making use of (9.8a) give:

$$(E_j^{(0)} - E_i^{(0)})\langle E_j^{(0)}\beta_k|E_i^{(0)}\rangle = 0, \text{ or } \langle E_j^{(0)}\beta_k|E_i^{(0)}\rangle = 0. \tag{9.9}$$

With degeneracy and (9.8a), the first order equation (9.5b) is still of the form:

$$(\hat{H}_0 - E_i^{(0)})|E_i^{(1)}\rangle + (\hat{H}_1 - E_i^{(1)})|E_i^{(0)}\rangle = 0.$$

$|E_i^{(0)}\rangle$ is, however, some unspecified linear combination of $|E_i^{(0)}\beta_j\rangle$ in view of (9.9):

$$|E_i^{(0)}\rangle = \sum_j \langle E_i^{(0)}\beta_j|E_i^{(0)}\rangle|E_i^{(0)}\beta_j\rangle. \tag{9.10}$$

The linear combination, or the expansion coefficients $\langle E_i^{(0)}\beta_j|E_i^{(0)}\rangle$, are yet to be determined. Multiplying (9.5b) by the bra-vector $\langle E_i^{(0)}\beta_{j'}|$ from the left and making use of (9.10) and the completeness theorem, $\hat{1} = \sum_{\beta_k}|E_i^{(0)}\beta_k\rangle\langle E_i^{(0)}\beta_k|$, give:

$$\sum_{\beta_k} \langle E_i^{(0)}\beta_{j'}|\hat{H}_1|E_i^{(0)}\beta_k\rangle\langle E_i^{(0)}\beta_k|E_i^{(0)}\rangle = E_i^{(1)}\langle E_i^{(0)}\beta_{j'}|E_i^{(0)}\rangle, \tag{9.11}$$

which means that to find the first order correction $E_i^{(1)}$ to the energy eigen value E_i and the zeroth-order eigen function $|E_i^{(0)}\rangle$, we need to diagonalize the matrix representing $\hat{H}_1$ within the manifold of degenerate states $|E_i^{(0)}\beta_j\rangle$. Depending upon the dimensionality of this matrix, this digonization procedure will yield a number of eigen values $E_{i\gamma}^{(1)}$ with the corresponding eigen function $|E_{i\gamma}^{(0)}\rangle$:

$$\sum_{\beta_k} \langle E_i^{(0)}\beta_{j'}|\hat{H}_1|E_i^{(0)}\beta_k\rangle\langle E_i^{(0)}\beta_k|E_{i\gamma}^{(0)}\rangle = E_{i\gamma}^{(1)}\langle E_i^{(0)}\beta_{j'}|E_{i\gamma}^{(0)}\rangle, \tag{9.11a}$$

which removes some of the degeneracy of the level $E_i^{(0)}$ in the absence of the perturbation. Note also that, because all these states are eigen functions of the unperturbed Hamiltonian corresponding to the eigen value $E_i^{(0)}$, to find $|E_i^{(0)}\rangle$ and $E_i^{(1)}$, the eigen value equation (9.11) is equivalent to:

$$\sum_{\beta_k} \langle E_i^{(0)}\beta_{j'}|\hat{H}|E_i^{(0)}\beta_k\rangle\langle E_i^{(0)}\beta_k|E_{i\gamma}^{(0)}\rangle = (E_i^{(0)} + E_{i\gamma}^{(1)})\langle E_i^{(0)}\beta_{j'}|E_{i\gamma}^{(0)}\rangle, \tag{9.11b}$$

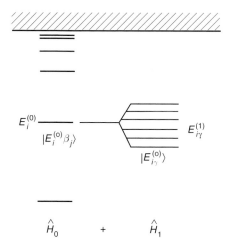

Figure 9.1. Schematic of the effect of a perturbation term in the Hamiltonian on the degenerate states

where the perturbed Hamiltonian $\hat{H}_1$ in (9.11) is replaced by the total Hamiltonian $\hat{H} \equiv \hat{H}_0 + \hat{H}_1$. The significance of this statement is not obvious at this point, but it will become clearer when applied to the molecular-orbital theory developed in the following Section, 9.2.

The contrast between the non-degenerate case and the degenerate case is that, in the former case, the first order effect of the perturbation leads to a *shift* of the unperturbed energy level. In the latter case, the unperturbed degenerate level is *split* into a number of new levels, $E_{i\gamma}^{(1)}$. The splittings between the new levels must be small compared to the separations between unperturbed levels for the degenerate perturbation theory to work.

In summary, in the degenerate case, the eigen values and the corresponding eigen functions are, respectively:

$$E_{i\gamma} = E_i^{(0)} + E_{i\gamma}^{(1)} + \dots, \tag{9.12a}$$

$$|E_{i\gamma}^{(0)}\rangle = \sum_{\beta_k} \langle E_i^{(0)} \beta_k | E_{i\gamma}^{(0)}\rangle |E_i^{(0)} \beta_k\rangle + \dots, \tag{9.12b}$$

where $E_{i\gamma}^{(1)}$ and $\langle E_i^{(0)} \beta_k | E_{i\gamma}^{(0)}\rangle$ are from the solutions of the eigen value equation (9.11a) or (9.11b). These results are shown schematically in Figure 9.1.

Although the theory presented in this section seems rather formal and formidable, in specific problems, it really is not very difficult to apply, as we will see in the following sections.

9.2 Covalent bonding of diatomic molecules

A collection of atoms will form a molecule, if it is energetically more favorable for them to do so than to exist as separate atoms. The quantum mechanic problem is then to compare the eigen energies of the atoms separately with when they exist together as a molecule.

Let us consider the simple case of a covalent bonded homo-nuclear diatomic molecule first. In the "molecular-orbital" approach, the general formulation of the problem is similar to that for the atoms. First, one looks for the eigen values and eigen states of the Hamiltonian for the single-electron states in the presence of the two nuclei separated by a distance R. These are the molecular-orbital states, which are analogous to the atomic orbitals in atoms with a single nucleus. The total number of electrons then occupy these available molecular-orbital states of successively higher energies according to Pauli's exclusion principle. The energy of the molecule is the sum of the energies of the electrons and the Coulomb repulsion of the ions in the molecule.

Consider first the electrons. Let the inter-atomic distance R be large enough for the atoms to be considered independent of each other initially. As they are brought together, the atoms will tend to form a molecule, if the total molecular energy decreases with decreasing R, and the molecule will stabilize at the inter-atomic distance R_m where its energy is at a minimum. We will demonstrate this qualitatively on the basis of the degenerate perturbation theory introduced in the previous section.

Let us assume that the molecule consists of two identical atoms A and B located at $x = \pm R_0/2$ (see Figure 9.2). The Hamiltonian for the single-electron states for the diatomic molecule is initially:

$$\hat{H}_0 = -\frac{\hbar^2}{2m}\nabla^2 + V(\vec{r}, R_0) = -\frac{\hbar^2}{2m}\nabla^2 - \frac{Ze^2}{\left|\vec{r} + \vec{R}_0/2\right|} - \frac{Ze^2}{\left|\vec{r} - \vec{R}_0/2\right|}. \tag{9.13}$$

Let us further assume that the two atoms are initially sufficiently far apart that the eigen functions centered on the two nuclei are essentially the atomic orbitals of the individual atoms $|A\rangle$ for $\left|\vec{r} + \vec{R}_0/2\right| \ll \left|\vec{r} - \vec{R}_0/2\right|$, and $|B\rangle$ for $\left|\vec{r} - \vec{R}_0/2\right| \ll \left|\vec{r} + \vec{R}_0\right|$, respectively, and the overlap between them is negligible, or $\langle A|B\rangle \approx 0$. If the atoms are far enough apart in the molecule initially, the single-electron molecular energy level $E_i = E_A = E_B$ is degenerate with two states: in one, $|A\rangle$, the electron is essentially at the atomic site A, and in the other, $|B\rangle$, the electron is essentially at the atomic site B. Thus, the initial degenerate eigen states and eigen value of the Hamiltonian (9.13) are, using the notations of the degenerate perturbation theory given in the previous section:

$$E_i^{(0)} = E_A = E_B, \tag{9.14a}$$

$$|E_i^{(0)}\beta_A\rangle \cong |A\rangle \text{ and } |E_i^{(0)}\beta_B\rangle \cong |B\rangle. \tag{9.14b}$$

Note that $E_i^{(0)}$ and $|E_i^{(0)}\beta_i\rangle$ refer to the unperturbed zeroth-order molecular states, while $|A\rangle$ and $|B\rangle$ refer to the unperturbed atomic states.

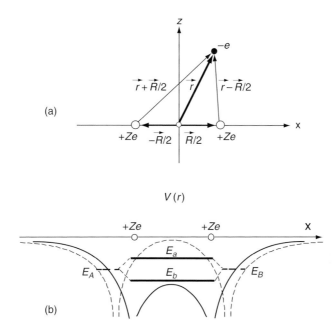

Figure 9.2. (a) A single electron in the skeleton of a homo-nuclear diatomic molecule. (b) Schematic showing the change in the Coulomb potential experienced by the electron in the molecular skeleton as the two nuclei are brought closer together (from the dashed to the solid curves). E_A and E_B correspond to the atomic orbitals; E_a and E_b correspond to the anti-bonding and bonding molecular orbitals, respectively. (see Eqs. (9.20a) and (9.20b), and Figure 9.3).

Let us now examine the effects on the energy of these states as the atoms are brought closer together and the Hamiltonian becomes:

$$\hat{H} = -\frac{\hbar^2}{2m}\nabla^2 + V(\vec{r}, R) = -\frac{\hbar^2}{2m}\nabla^2 - \frac{Ze^2}{|\vec{r} + \vec{R}/2|} - \frac{Ze^2}{|\vec{r} - \vec{R}/2|}, \tag{9.15}$$

where $R < R_0$. We assume that the change in R and, hence, in the potential terms in the Hamiltonian is small enough that the degenerate perturbation theory developed in the previous section applies. It is also assumed that R is finite and sufficiently large that the two atomic orbitals $|A\rangle$ and $|B\rangle$ are still approximately "orthogonal" in the sense that $\langle A|B\rangle \approx 0$. *These are drastic approximations that are only good enough to give a qualitative indication of the bonding mechanism between the atoms, and the model is not adequate to yield any serious quantitative results.* Nevertheless, the perturbed eigen values and eigen functions can be found by diagonalizing the matrix representing the Hamiltonian, (9.15), within the two degenerate states given in (9.14b). The new molecular eigen states are of the form:

$$|E_{i\gamma}^{(0)}\rangle = C_{A\gamma}|A\rangle + C_{B\gamma}|B\rangle, \tag{9.16}$$

where $\left|C_{A\gamma}\right|^2 + \left|C_{B\gamma}\right|^2 = 1$. The 2×2 matrix equation corresponding to (9.11b) is:

$$\begin{pmatrix} \langle A|\hat{H}|A\rangle & \langle A|\hat{H}|B\rangle \\ \langle B|\hat{H}|A\rangle & \langle B|\hat{H}|B\rangle \end{pmatrix} \begin{pmatrix} C_{A\gamma} \\ C_{B\gamma} \end{pmatrix} = \left(E_A + E_{i\gamma}^{(1)} \right) \begin{pmatrix} C_{A\gamma} \\ C_{B\gamma} \end{pmatrix}. \tag{9.17}$$

Because of the spatial symmetry of the Hamiltonian and $|A\rangle$ and $|B\rangle$ under inversion, $\vec{r} \rightarrow -\vec{r}$, the diagonal and off-diagonal elements of the matrix are equal:

$$\langle A|\hat{H}|A\rangle = \langle B|\hat{H}|B\rangle \equiv \bar{E}_i \tag{9.18a}$$

$$\langle A|\hat{H}|B\rangle = \langle B|\hat{H}|A\rangle. \tag{9.18b}$$

Thus, solving Eq. (9.17) yields two new eigen values corresponding to a bonding and an anti-bonding state ($-$ and $+$ signs, respectively, below):

$$E_{i\gamma} = E_A + E_{i\gamma}^{(1)} = \bar{E}_i \pm \left|\langle A|\hat{H}|B\rangle\right| = \bar{E}_i \pm \left|\langle A|\Delta\hat{H}|B\rangle\right|, \tag{9.19}$$

where $\Delta\hat{H} = \hat{H} - \hat{H}_0 = \hat{V}(\vec{r}, R) - \hat{V}(\vec{r}, R_0) \equiv \Delta\hat{V}$ gives the change in the Coulomb potential (see Figure 9.2) experienced by the electron due to the change in R from R_0 as the atoms are brought closer to each and assuming that $\langle A|\hat{H}_0|B\rangle = E_i^{(0)}\langle A|B\rangle \approx 0$. The initially degenerate level is now split into two levels with the splitting between the two equal to $2\left|\langle A|\Delta\hat{H}|B\rangle\right|$, and there is a slight down shift of the average of the two energy levels $\bar{E}_i$ from E_A (note that $\langle A|\Delta\hat{H}|A\rangle < 0$). The two split levels, (9.19), represent the "bonding" and "anti-bonding" states of the molecule with the eigen energies:

$$E_{ib}^{(1)} = \bar{E}_i - \left|\langle A|\Delta\hat{H}|B\rangle\right| \tag{9.20a}$$

and

$$E_{ia}^{(1)} = \bar{E}_i + \left|\langle A|\Delta\hat{H}|B\rangle\right|, \tag{9.20b}$$

respectively. Solution of (9.17) also gives the corresponding bonding and anti-bonding orbitals:

$$\left|E_{ib}^{(0)}\right\rangle = C_{Ab}|A\rangle + C_{Bb}|B\rangle \tag{9.21a}$$

and

$$\left|E_{ia}^{(0)}\right\rangle = C_{Aa}|A\rangle + C_{Ba}|B\rangle, \tag{9.21b}$$

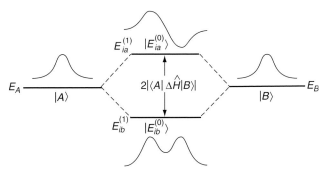

Figure 9.3. Bonding and anti-bonding orbitals of the homo-nuclear diatomic molecule and the original atomic orbitals of the constituent atoms. (For clarity, the slight shift of $\bar{E}_i$ from E_A is neglected in this schematic diagram.)

where

$$\frac{C_{Ab}}{C_{Bb}} = -\frac{\langle A|\Delta\hat{H}_1|B\rangle}{|\langle A|\Delta\hat{H}_1|B\rangle|} = +1, \tag{9.21c}$$

$$\frac{C_{Aa}}{C_{Ba}} = +\frac{\langle A|\Delta\hat{H}_1|B\rangle}{|\langle A|\Delta\hat{H}_1|B\rangle|} = -1. \tag{9.21d}$$

Qualitatively, the reason that $\langle A|\Delta\hat{H}_1|B\rangle = -|\langle\Delta\hat{H}_1|B\rangle|$ is that, as R decreases from R_0, the potential energy decreases over the region where the overlap between the atomic orbitals $|A\rangle$ and $|B\rangle$ is the largest, or $\Delta\hat{H}_1 < 0$ in this region (see Figure 9.2). The normalized bonding and anti-bonding molecular orbitals, (9.21a) and (9.21b), of the homo-nuclear diatomic molecule are, therefore, symmetric and anti-symmetric combinations of the atomic orbitals, respectively:

$$
\begin{aligned}
|E_{ib}^{(0)}\rangle &= \frac{1}{\sqrt{2+2S}}[|A\rangle + |B\rangle] \\
|E_{ia}^{(0)}\rangle &= \frac{1}{\sqrt{2+2S}}[|A\rangle - |B\rangle],
\end{aligned}
\tag{9.22}
$$

where $S = \langle A|B\rangle \approx 0$ is the overlap integral of the two atomic orbitals and assumed negligibly small. These results are shown schematically in Figure 9.3. The molecule will stabilize at $R = R_m$, where the total molecular energy, including the attractive covalent bonding energy of the electrons and the Coulomb repulsive energy of the ions, is at a minimum.

If each of the two atoms has only one valence electron, the two electrons will both occupy the bonding orbital with opposite spins. For example, when two hydrogen atoms in the ground 1s level are brought together, the energy of the single-electron bonding state is reduced by about 2.7 eV from the ground state energy of −13.6 eV in an isolated hydrogen atom with R_m stabilized around 1.1 Å. Introducing a second electron with the opposite spin into the ground state of the hydrogen molecule will increase the binding energy of the molecule further to about 4.47 eV with an inter-nuclear distance of ~0.7 Å. Thus, the two atoms can form a stable diatomic hydrogen molecule with both electrons in the bonding molecular orbital formed from the 1s atomic orbitals.

The perturbation theory given here is adequate to show how the two atoms can form a diatomic molecule through the covalent bonding mechanism. To show that the diatomic molecule will stabilize around an equilibrium inter-atomic distance R_m, one has to evaluate $\langle E_{i\gamma}^{(0)}|\hat{H}|E_{i\gamma}^{(0)}\rangle$ against variations in $C_{A\gamma}$ and $C_{B\gamma}$, taking into account also the Coulomb repulsion between the ions, and find at what R the total molecular energy is a minimum. This is the basis of a rigorous molecular-orbital theory (see, for example, Ballhausen and Gray (1964); Coulson (1961)), which is beyond the scope of this book.

For other multi-electron atoms in a diatomic molecule, the electrons will fill the available molecular orbital states of successively higher energies according to Pauli's exclusion principle, and the molecular wave functions are the appropriate Slater determinants just like in the multi-electron atoms discussed in Chapter 7. This kind of bonding mechanism between two atoms is called "covalent bonding," where the atoms share the valence electrons and bond to form a molecule. The same considerations can be extended to hetero-nuclear diatomic and to poly-atomic molecules. The details are, of course, more complicated, but the principles are the same.

9.3   sp, sp^2 and sp^3 orbitals and examples of simple organic molecules

If the valence states of the separated atoms forming the molecular states have orbital degeneracy, each such atom can form multiple bonds with another atom or other atoms. Consider, for example, such a multi-electron atom as carbon. Carbon is of particular interest, because it is the basic element in organic chemistry, and the crystalline structure of carbon in the form of diamond crystal is the same as that of such important semiconductors as Si and Ge. The diamond structure is also closely related to the zincblende structure of such important compound semiconductors as GaAs and ZnS. From Table 7.1, it is seen that carbon has a total of six electrons. The ground-state configuration is $1s^2 2s^2 2p^2$. The valence states are the 2s and 2p states. The p states are defined in (6.24) and Table 6.1 and have three-fold orbital degeneracy corresponding to the orthogonal orientations of the orbitals, as shown in Figure 6.5(b) and (c).

When the carbon atom forms a bond with other atoms, such as hydrogen or another carbon, the valence states that go into forming the bonding orbital with the lowest molecular energy can often be linear combinations of the 2s state and the three 2p states. This is because the shift in energy between the 2s and 2p states of the carbon atom can be smaller than or comparable to the interaction energy between the s and p orbitals of the atoms forming the bond. In applying the perturbation theory, the manifold of degenerate states must be expanded to include the near-degenerate atomic states, and the basis states used are linear combinations of the expanded basis. This is

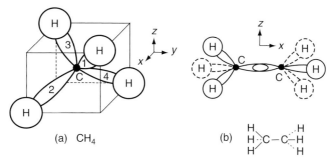

Figure 9.4. Schematics showing four hybridized sp^3 orbitals of carbon covalent bonded (a) to four hydrogen 1s orbitals to form a methane molecule, and (b) to three hydrogen atoms and a –CH$_3$ radical group to form an ethane molecule.

often the case when ns and np states are the valence states and the principal quantum number n is not too large, such as in carbon ($n = 2$), Si ($n = 3$), and Ge ($n = 4$). This process of forming mixed states of the atomic orbitals in the process of forming the molecular orbitals is called "hybridization." One 2s and three 2p states can form two, three, or four hybridized states in various combinations, depending on the molecular complex the carbon atom goes into. We consider first some simple organic molecules involving hybridized sp^n states:

sp^3 orbitals

Methane

Consider, for example, CH$_4$, the methane molecule, consisting of one carbon atom covalent-bonded to four hydrogen atoms. The four normalized hybridized sp^3 orbitals of carbon are:

$$|1\rangle = \frac{1}{2}\left[|s\rangle + |p_x\rangle + |p_y\rangle + |p_z\rangle\right], \tag{9.23a}$$

$$|2\rangle = \frac{1}{2}\left[|s\rangle + |p_x\rangle - |p_y\rangle - |p_z\rangle\right], \tag{9.23b}$$

$$|3\rangle = \frac{1}{2}\left[|s\rangle - |p_x\rangle - |p_y\rangle + |p_z\rangle\right], \tag{9.23c}$$

$$|4\rangle = \frac{1}{2}\left[|s\rangle - |p_x\rangle + |p_y\rangle - |p_z\rangle\right]. \tag{9.23d}$$

They are shown schematically in Figure 9.4(a). Each of these orbitals can form a hetero-nuclear diatomic covalent bond with a hydrogen atom to form a methane molecule, as shown in Figure 9.4(a). As can be calculated easily from this figure, the angle between these bonds based on this simple model should be 109.47°.

Ethane

The four hybridized sp^3 orbitals (9.23a–d) do not all have to be bonded to the same kind of atoms. One of these can be replaced by a "radical" group CH_3 to form a new molecule, in this case an ethane molecule CH_3–CH_3, as shown in Figure 9.4(b). The angles between the C–H bonds and between the C–H and the C–C bonds are all $109.47°$.

sp^2 Orbitals

Ethylene

Similarly, the s orbital can hybridize with, for example, the p_x and p_z orbitals to form three sp^2 orbitals:

$$|x_\pm \rangle = \frac{1}{\sqrt{3}} \left[|s\rangle \pm \sqrt{2}|p_x \rangle \right], \tag{9.24a}$$

$$|2\pi/3\rangle = \frac{1}{\sqrt{3}} \left[|s\rangle \mp \frac{1}{\sqrt{2}}|p_x\rangle + \sqrt{\frac{3}{2}}|p_z \rangle \right], \tag{9.24b}$$

$$|-2\pi/3 \rangle = \frac{1}{\sqrt{3}} \left[|s\rangle \mp \frac{1}{\sqrt{2}}|p_x\rangle - \sqrt{\frac{3}{2}}|p_z\rangle \right]. \tag{9.24c}$$

These orbitals can bond with two hydrogen atoms in one x direction and another similar carbon atom in the opposite x direction. The angle between the sp^2 bonds is $120°$ as indicated. With the additional p_y orbitals, there can be a double-bond between the two carbon atoms that are each attached to two hydrogen atoms to form an ethylene molecule $H_2C = CH_2$, as shown in Figure 9.5(a).

sp orbitals

Acetylene

The s orbital can also hybridize with a single p_x orbital to form two sp -orbitals pointed in the +x and −x directions as in the acetylene molecule:

$$|x_+\rangle = \frac{1}{\sqrt{2}} \left[|s\rangle + |p_x\rangle \right] \tag{9.25a}$$

$$|x_-\rangle = \frac{1}{\sqrt{2}} \left[|s\rangle - |p_x\rangle \right]. \tag{9.25b}$$

One of these forms a bond with a hydrogen atom on one end (say, $−x$ direction) and with another similar carbon atom in the other end (+x direction), which is similarly

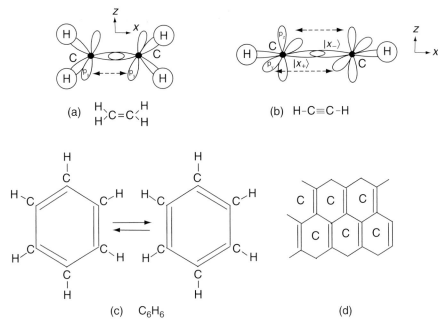

Figure 9.5. Schematics of the (a) ethylene, (b) acetylene, (c) benzene molecules (*Kikulé* structures), and (d) graphite.

bonded with another hydrogen atom, as shown in Figure 9.5(b). The remaining p_z and p_y orbitals of the two carbon atoms then form two additional covalent bonds between the carbon atoms. Thus, the carbon–carbon bond is a triple-bond, while the remaining bond of each carbon atom bonds to a hydrogen atom and forms an $H-C \equiv C-H$ molecule, which is the acetylene molecule.

Benzene ring structures

The carbon atoms do not have to form linear structures only. With suitably hybridized and oriented sp^2 and p orbitals of carbon, six carbon atoms and six hydrogen atoms can be brought together to form a benzene molecule, C_6H_6, in a ring structure, as shown in Figure 9.5(c). The problem is actually more complicated because there are, for example, two equivalent structures with the same energy, as shown in this figure. In this case, there is a 50–50 probability that each C–C bond is a single- or a double-bond, as shown. In the language of the chemists, the "resonance" between these two so-called "*Kikulé* structures" leads to additional stabilization of the molecule.

The benzene ring structure is the basic building block of a great variety of organic and inorganic molecules and solids. For example, the carbon ring does not have to bond with hydrogen atoms only. It can bond with six other carbon ring structures that further connect with other carbon rings *ad infinitum* and form a gigantic sheet, which is the structure of graphite. The planar structure of graphite accounts for its superior property as a lubricant.

9.4 Diamond and zincblende structures and space lattices

In addition to the linear and planar structures, a three-dimensional crystalline structure, the diamond structure, can also be constructed from the tetrahedral complexes of carbon through its hybridized sp^3 orbitals. This is an exceedingly important structure for electronics and photonics, for such important IV–IV semiconductors as the Si and Ge crystals have the same structure. In addition, the zincblende structure of III–V binary semiconductors and some of the II–VI compounds is closely related to the diamond structure.

The basic tetrahedral complex of the carbon atoms is shown in Figure 9.6(a). It is similar to the methane molecule shown in Figure 9.4(a), except that the hydrogen atoms are replaced by other similar carbon atoms. The four sp^3 orbitals are given in Eq. (9.23a–d). Each carbon atom can thus be bonded to four other carbon atoms through the four hybridized sp^3 orbitals to form the tetrahedral complex. Each tetrahedral

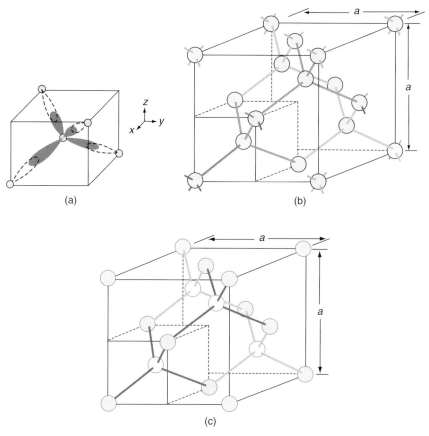

Figure 9.6. (a) The tetrahedral complex of the sp^3 orbitals of carbon, and (b) the diamond structure of carbon, silicon, and germanium crystals. (c) The zincblende structure, which is similar to the diamond structure except each atom is bonded to four atoms of a different kind (e.g. Ga and As, forming the binary semiconductor GaAs crystal).

complex can be bonded to four other similar complexes, as shown in Figure 9.6(b). Extending these tetrahedral complexes throughout three-dimensional space leads to a space lattice of "diamond structure," which is the basic structure of, for example, covalent-bonded IV–IV crystals such as diamond (carbon), silicon, and germanium crystals.

If the centers of the tetrahedral complex shown in Figure 9.6(a) are replaced by atoms from column III (or V) of the periodic table (see, for example, Table 7.1) while the corners are replaced by column V (or III) atoms as shown in Figure 9.6(c), the resulting crystalline structure is the zincblende structure. In this case, there is some migration of negative charge from the V-atom to the III-atom for each bond. The bonding is then partially covalent and partially ionic or electrostatic. Many III–V compounds, such as GaAs, GaP, GaN, InAs, InP, and InSb, are semiconductors of great practical importance. Some of the II–VI compounds, such as ZnS, ZnSe, CdS, and CdSe, can also form space lattices of zincblende structure with partial covalent bonding and still larger ionicity than the bonds between III–V atoms, but some of these II–VI crystals can have both cubic symmetry or hexagonal symmetry.

A crystalline solid is also like a giant multi-electron molecule. Depending on how tightly the valence electrons are bound to the atoms, the electronic properties of the solid can be better understood on the basis of either a "nearly-free-electron model" or a "tight-binding model." In either case, it is based on the basic ideas of time-independent perturbation theory, as outlined in Section 9.1. In the case of the tight-binding model, the starting point is the individual atoms. This model is more suited for insulators and wide band-gap semiconductors. The interaction of the atoms with its neighbors is considered a small perturbation on the atomic states. In the nearly-free-electron model, which is more suited for metals and narrower band-gap semiconductors, the solid is considered a giant quantum well of macroscopic dimensions. The potential for the valence electrons inside the well is *almost* spatially independent, and the valence electrons themselves are delocalized and belong to the entire solid. The spatially fixed periodic potential due to the lattice ions is considered a perturbation that modifies the free-electron states, leading to the "Bloch states" in the well. For applications in semiconductor electronics and photonics, the nearly-free-electron model based on the Bloch theorem is the commonly used approach. It will be discussed in more detail in the next chapter.

9.5 Problems

9.1 Consider the spin–orbit interaction term for hydrogen of the form (6.62). Write the matrix corresponding to this term in the six-fold degenerate states with the same orbital angular momentum quantum number $\ell = 1$ in the representation in which $\hat{L}^2$, $\hat{L}_z$, $\hat{S}^2$, $\hat{S}_z$ are diagonal. Diagonalize this matrix according to the degenerate perturbation theory and find the corresponding eigen values and eigen functions. Compare the eigen values obtained with the corresponding results (the original degenerate states split into two new degenerate levels: shifted by $\xi_{n\ell}/2$ and $-\xi_{n\ell}$ corresponding to $j = 3/2$, $m_j = \pm 3/2$, $\pm 1/2$ and $j = 1/2$,

$m_j = \pm 1/2$) given in Section 6.5. The eigen functions give the relevant vector-coupling coefficients $\langle \ell m_\ell s m_s | j m_j \ell s \rangle$ defined in (6.59) for this particular case. (Hint: the 6×6 matrix corresponding to the manifold of degenerate states to be diagonalized breaks down to two 2×2 and two 1×1 matrices down the diagonal by suitable ordering of the rows and columns of matrix elements. The smaller 2×2 matrices can then be diagonalized easily.)

9.2 Extend the perturbation theory for the covalent bonded homo-nuclear diatomic molecule to the case of hetero-nuclear diatomic molecules. More specifically, find the energies and the corresponding wave functions of the bonding and anti-bonding orbitals of the molecule in terms of the energies of the atoms E_A and E_B, where $E_A \neq E_B$, and the corresponding wave functions $|\Psi_A\rangle$ and $|\Psi_B\rangle$, respectively.

9.3 Suppose the un-normalized molecular orbital of a diatomic homo-nuclear diatomic molecule is:

$$|\Psi_{\text{mo}}\rangle = C_A |A\rangle + C_B |B\rangle,$$

where $|A\rangle$ and $|B\rangle$ are the normalized atomic orbitals.

(a) Normalize the above molecular orbital.

(b) Find the energies and wave functions for the bonding and anti-bonding molecular states by minimizing the energy $E = \langle \Psi_{\text{mo}} | \hat{H} | \Psi_{\text{mo}} \rangle$, where $|\Psi_{\text{mo}}\rangle$ is the normalized molecular orbital, against variations in the relative contributions of the atomic orbitals making up the normalized molecular orbital in the limit of negligibly small overlap integral between the atomic orbitals ≈ 0. (Hint: solve for E from the secular equation by setting $\frac{\partial E}{\partial C_A} = 0$ and $\frac{\partial E}{\partial C_B} = 0$; then find C_A and C_B.) Compare the resulting energy values and the corresponding wave functions with the bonding and anti-bonding energies (9.20a and b) and wave functions (9.21a–d), respectively, on the basis of the perturbation theory outlined in the text.

9.4 Consider the diamond lattice shown in Figure 9.6(b). Find the number of atoms per cube cell of the volume a^3 in such a lattice. What is the number of valence electrons per such a unit cell (the cubic cell shown in Figure 9.6(b) or (c) is termed a "conventional unit cell" in contrast to the "primitive unit cell") for the diamond crystal and for the silicon crystal?

9.5 The primitive translational vectors $\vec{a}$, $\vec{b}$, and $\vec{c}$ of a periodic lattice are defined by the equation: $\vec{R} = n_1 \vec{a} + n_2 \vec{b} + n_3 \vec{c}$, where $\vec{R}$ is the displacement vector connecting any two lattice points in the periodic lattice and n_1, n_2, and n_3 are integers 1, 2, 3, … Find the primitive translational vectors of a simple cubic lattice (repeated simple cubes with lattice points at the corners of the cubes) and of a face-centered cubic lattice (repeated cubes with lattice points at the corners of the cubes and the centers of the faces).

9.6 Show that the diamond lattice is simply two interlaced face-centered cubic latticed displaced one quarter of the length along the diagonal of the cube.

9.7 What is the length of the C–C bond in the diamond lattice expressed as a fraction of cubic edge "a" shown in Figure 9.6(b)?

10 Electronic properties of semiconductors and the p–n junction

Some of the most important applications of quantum mechanics are in semiconductor physics and technology based on the properties of electrons in a periodic lattice of ions. This problem is discussed on the basis of the nearly-free-electron model of the crystalline solids in this chapter. In this model, the entire solid is represented by a quantum well of macroscopic dimensions. The spatially-varying electron potential due to the periodic lattice of ions inside the well is considered a perturbation on the free-electron states leading to the Bloch states and the band structure of the semiconductor. The concepts of effective mass and group velocity of the electrons and holes in the conduction and valence bands separated by an energy-gap are introduced. The electrons and holes are distributed over the available Bloch states in these bands depending on the location of the Fermi level according to Fermi statistics. The transport properties of these charge-carriers and their influence on the electrical conductivity of the semiconductor are discussed. When impurities are present, the electrical properties can be drastically altered, resulting in n-type and p-type semiconductors. The p–n junction is a key element in modern semiconductor electronic and photonic devices.

10.1 Molecular orbital picture of the valence and conduction bands of semiconductors

Atoms can be brought together to form crystalline solids through a variety of mechanisms. Most of the commonly used semiconductors are partially covalently and partially ionically bonded crystals of diamond or zincblende structure. For the column IV elements, each atom starts out with exactly four valence electrons (s^2p^2) occupying two s and two p spin-degenerate atomic orbital states. In the covalent bonded solids, the s and p atomic orbitals are hybridized and form four sp^3 orbitals attached to each atomic site, as shown in Figure 9.6(a). Each bond has two spin states and can accommodate two electrons. In the ground state of the solid, each Group IV atom contributes one electron to fill the two available spin states of each diatomic bond; all the available bonding states are, thus, filled exactly by the available valence electrons from each atom. If one of these electrons is excited into an anti-bonding sp^3 state, it will leave a hole on the bond. In the crystalline solid, every electron is indistinguishable from every other and every site is indistinguishable from every other equivalent site.

Thus, the electron states and hole states are not localized on any particular bond but are linear combinations of the bonding and anti-bonding states of all the bonds that are the eigen states of the whole crystal. These states are broadened because of the interactions among the bonds. The bonding states in the IV–IV semiconductors, for example, form the "valence band" which is fully occupied in the ground state of the solid. The anti-bonding states form the "conduction band." It is completely empty when the solid is in the ground state and the valence band is full. When an electron is excited, it will occupy one of the conduction band states of the whole crystal. Since there are many other conduction band states which the excited electron can move to, it can lead to electric current flow in the solid – hence the name "conduction band."

In solids in general, if the gap between the valence and conduction bands is much greater than the thermal energy of the electrons, there are very few electrons in the conduction band of the solid; it is, thus, an insulator. If the gap is relatively small, on the order of 1 eV, for example, it is a semiconductor. In the limit of no gap, it is a metal. The "band structure" of the crystal is, therefore, clearly of fundamental importance in determining its electrical characteristics. In this section, we will develop a qualitative picture of the solid based on a qualitative molecular-orbital picture first. This will be followed by a more formal and rigorous formalism based on the Bloch states in the following section.

There are two possible ways to view the problem of how the valence band and the conduction band in a semiconductor may arise from, for example, the s and p orbitals of its constituent atoms. They reflect different ways of applying the time-independent perturbation theory to the problem.

In one version, it is very much like what happens in the diatomic molecule discussed in Section 9.2. In the solid, suppose there are a large number of atoms per unit volume (maybe $10^{23} \, \text{cm}^{-3}$). If there is no interaction between any of the atoms, then the single-electron energy levels E_s and E_p of the solid are highly degenerate. When the atoms are brought together to form a covalent bonded solid, the neighboring atoms will interact with each other and form diatomic bonds, each with a bonding and an anti-bonding molecular state. Because some of the degenerate atomic p orbitals pointing in the direction of the bond are spatially more extended along the bond direction than the other orbitals, the overlap between these p orbitals is larger than those between the other orbitals. The split between the corresponding anti-bonding and bonding states is, therefore, larger than those between some of the other p and the s orbitals, and may even be larger than the shift between the atomic energy levels E_s and E_p, as shown in Figure 10.1(a)., If there are interactions between the bonds, these molecular states will become more delocalized and there will be additional broadening into bands, as shown in Figure 10.1(b). There may be mixing of the bonding-states formed from the p orbitals and the s orbitals, leading to the formation of the valence band of the solid with the top of the valence band most probably p-like. The mixing of the s and p orbitals in each band is analogous to the hybridization of the s and p orbitals in forming the covalent bonds in diatomic molecules, as discussed in Section 9.2. The broadened anti-bonding states will likewise form the conduction band of the solid with the bottom of the band most probably s-like. In the case of the IV–IV compounds in

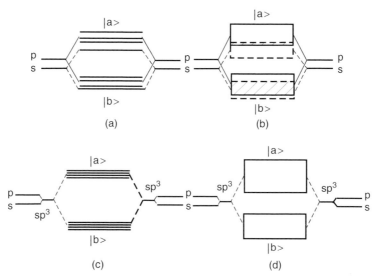

Figure 10.1 Schematics showing qualitatively the parentage of the energy eigen states of the sp^3 bonded crystal. (a) Bonding and anti-bonding states formed from the atomic s and p orbitals with no bond interaction. (b) Broadening of the bonding and anti-bonding states of (a) due to bond interactions. The molecular states originated from the atomic p orbitals are framed approximately by solid lines; those from the s orbitals are framed approximately by the dashed lines. The hatched regions indicate where there is appreciable mixing of these states. (c) Bonding and anti-bonding states of the sp^3 hybridized orbital with no bond interaction. (d) Broadening of the bonding and anti-bonding states in (c) due to bond interactions. (See the text for additional explanations.)

the ground states, the four electrons from each column IV atom will exactly fill the available valence band states formed from the bonding orbitals.

In the second view, the s and p orbitals are hybridized first and then form bonding and anti-bonding states of the bonds, as shown in Figure 10.1(c). Again, if there are interactions between pairs of bonds, the molecular states of the bonds will delocalize and broaden into a valence band of lower energy and a conduction band of higher energy with possibly a gap in between, as shown in Figure 10.1(d).

These simple pictures do not show, however, how the energies of the electrons and holes in the solid vary with the linear momentum of the particles. For this, we need to have the variation of the energy in the wave vector $\overrightarrow{k}$-space of the de Broglie waves corresponding to the particles in the periodic lattice. It will come from a more rigorous description of the eigen states of the Hamiltonian of the single-electron states of the whole crystal based on the Bloch theorem, to be described in the next section.

10.2 Nearly-free-electron model of solids and the Bloch theorem

In the nearly-free-electron model, the crystal is represented by a quantum well of macroscopic dimensions. The Coulomb potentials between the atomic sites are

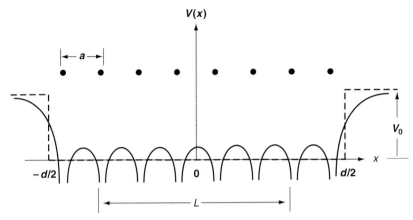

Figure 10.2 Schematic of a linear array of ion cores (solid dots) and the corresponding periodic crystal potential (solid curves) and the quantum well (dashed lines) model.

reduced from that of the individual atoms due to the opposing fields of the ion cores of the atoms in the solid. This reduction of the Coulomb potential between the ions can cause the atomic orbitals to mix with those of their neighbors and lead to broadening in energy and in the spatial extent of the electron charge distribution. In the case of metals, this can even free the valence electrons from the atoms and allow them to roam freely in the whole solid. In semiconductors, enough electrons can be freed from the valence band at the operating temperature of the solid and be excited into the conduction band to drastically alter the electrical characteristics of the solid.

Consider, for example, a one-dimensional linear periodic array of atoms with the electron potential energy due to the ion cores inside the crystal, as shown schematically in Figure 10.2:

$$V(x) = \begin{cases} V_0, & \text{for} & x < -d/2, \\ V_{\text{cr}}(x), & \text{for} & -d/2 < x < +d/2, \\ V_0, & \text{for} & x > d/2, \end{cases} \qquad (10.1)$$

where the crystal potential has the translational-symmetry property:

$$V_{\text{cr}}(x + a) = V_{\text{cr}}(x) \qquad (10.2)$$

and a is the periodicity of the lattice. The corresponding time-independent Schrödinger equation is, for $-d/2 < x < +d/2$:

$$\hat{H}\Psi_E(x) \equiv \left[\hat{H}_0 + V_{\text{cr}}(x)\right]\Psi_E(x)$$

$$= \left[-\frac{\hbar^2}{2m}\frac{\partial^2}{\partial x^2} + V_{\text{cr}}(x)\right]\Psi_E(x)$$

$$= E\Psi_E(x). \qquad (10.3)$$

Physically, it is clear that, because the crystal is invariant under the translation $x \to x + a$, except near the edges, the charge distribution in the crystal must also have the same translational-invariance property, or:

$$|\Psi_E(x + a)|^2 = |\Psi_E(x)|^2 \tag{10.4}$$

for all values of x. Thus, the wave function itself can differ from a purely periodic function by at most a phase factor, and must be of the form:

$$\Psi_{E(k)}(x) \equiv u_{E(k)}(x)e^{ikx}, \tag{10.5}$$

where

$$u_{E(k)}(x + a) = u_{E(k)}(x) \tag{10.6}$$

is periodic with the periodicity a . The free-particle wave function e^{ikx} of the overall wave function $\Psi_{E(k)}(x)$ is sometimes called its "envelope function." Note that E will now depend on the value of k . Because of the periodic condition, $u_{E(k)}(x)$ can also be expanded as a Fourier series of the form:

$$u_{E(k)}(x) = \sum_{n = 0, \pm 1, \pm 2, \pm 3, \ldots} C_n(k)e^{iG_n x}, \tag{10.6a}$$

where $G_n = \dfrac{n \cdot 2\pi}{a}$. This is in essence the Bloch theorem, which states that: "the eigen functions of the time-independent Schrödinger equation with a periodic potential are of the form (10.5) and (10.6) or (10.6a)." As shown in Chapter 3, an electron with a fixed linear momentum p_x in free space is a de Broglie wave with a wave number $k = \dfrac{p_x}{\hbar} = \dfrac{\sqrt{2mE}}{\hbar}$ and a constant amplitude. From Bloch's theorem, the de Broglie wave of an electron in a periodic potential well region is a spatially amplitude-modulated wave with a periodicity equal to the lattice spacing of the periodic structure and a "crystal momentum" of $\hbar k$. The eigen functions and the corresponding eigen values now depend on the wave number k:

$$\left[-\frac{\hbar^2}{2m}\frac{\partial^2}{\partial x^2} + V_{cr}(x) \right] \Psi_{E(k)}(x) = E(k)\,\Psi_{E(k)}(x), \tag{10.7}$$

and

$$\Psi_{E(k)}(x) = u_{E(k)}(x)e^{ikx} = \sum_{n = 0, \pm 1, \pm 2, \pm 3, \ldots} C_n(k)e^{i(k + G_n)x}. \tag{10.7a}$$

The allowed values of k are determined by the boundary conditions on the overall wave function $\Psi_{E(k)}(x)$. Since the interest here is in the intrinsic property of the material, we consider a large uniform section of the crystal from $x = -L/2$ to $+L/2$ spanning over a large number of lattice sites in an infinitely large crystal ($d \to \infty$ in

Fig 10.2). For a large uniform crystal, a commonly used boundary condition on the overall wave function is the cyclic boundary condition of Born and Von Karman (see, for example, Cohen-Tannoudji *et al.* (1977) Vol. II, p. 1441):

$$\Psi_{E(k)}(-L/2) = \Psi_{E(k)}(+L/2). \tag{10.8}$$

L can be chosen to be an exact integral multiple of the lattice spacing a so that:

$$u_{E(k)}(x = -L/2) = u_{E(k)}(x = +L/2). \tag{10.8a}$$

Thus, from (10.7a), (10.8), and (10.8a), the envelope function e^{ikx} of the corresponding amplitude-modulated de Broglie wave at $x = -L/2$ must be the same as it is at $x = L/2$, or:

$$e^{-ikL/2} = e^{ikL/2}$$

or

$$e^{ikL} = 1,$$

and the allowed values of k must be:

$$k = \pm\frac{2N\pi}{L}, \quad \text{where } N = 1, 2, 3, 4, \ldots \tag{10.9}$$

If it is a finite section of the crystal in, for example, some quantum well structure, then the specific boundary conditions on the wave function at the surfaces ($x = \pm L/2$) of the section of the crystal must be taken into account in the boundary conditions on the envelope function. Otherwise, for the cyclic boundary condition case, (10.9), it means that there are an integral number N of de Broglie wavelengths, $2\pi/k = \lambda_d$, in the length L. For L very large, k becomes a continuum. Note that, in k-space, the number of allowed k-values between $-\pi/a$ and $+\pi/a$ is exactly equal to the number, L/a, of lattice sites separated by a within the length L of the spatially uniform crystal. The range in k-space between $-\pi/a$ and $+\pi/a$ is called the first "Brillouin zone" and the points $\pm\pi/a$ are the corresponding "zone boundaries." Repeating this, the entire k-space can be divided up into Brillouin zones. The range between $\pm(N-1)\pi/a$ and $\pm N\pi/a$ forms the Nth Brillouin zone and $\pm N\pi/a$ defines the boundaries of the Nth Brillouin zone. The eigen states near the zone boundaries are of special importance in the electronic properties of the semiconductors, as will be shown later.

For a general three-dimensional periodic lattice, the crystal potential has the "translational-invariance" property:

$$V_{cr}(\vec{r}) = V_{cr}(\vec{r}' + \vec{R}),$$

where

$$\vec{R} = n_1\vec{a} + n_2\vec{b} + n_3\vec{c}$$

is the vector connecting any two lattice points $\vec{r}$ and $\vec{r}'$ in the crystal. n_1, n_2 and n_3 are integers. $\vec{a}$, $\vec{b}$, and $\vec{c}$ are the "primitive translational vectors," or a set of three independent shortest vectors connecting two lattice points that define the three-dimensional lattice. For three dimensions, the concept of Bloch states, (10.5)–(10.6a), must be generalized accordingly. Note also that, in 3-D structures, the choice of the primitive translational vectors for any lattice structure is not unique – as long as repeating the primitive translational vectors $\vec{a}$, $\vec{b}$, and $\vec{c}$ can, and must, generate all the lattice points in the lattice structure. The three primitive translational vectors form a "primitive unit cell" of the crystalline structure. Repeating the primitive unit cells must fill the entire crystalline space. Thus, the number of valence electrons per volume of the crystal can be determined from the number of atoms per primitive unit cell and the number of valence electrons per atom. The number of valence electrons per volume will in turn determine the electrical properties of the crystal, be it metal, insulator, or semiconductor, as we shall see below.

10.3 The *k*-space and the *E* vs. *k* diagram

Returning now to the simpler one-dimensional case again, because the crystal potential is periodic in x with a period, or a "primitive translation," a,

$$V_{\mathrm{cr}}(x + a) = V_{\mathrm{cr}}(x).$$

As a periodic function in x, it can be put in the form of a spatial Fourier series:

$$V_{\mathrm{cr}}(x) = \sum_{n = \pm1, \pm2, \pm3, \ldots} V_n e^{iG_n x}, \tag{10.10}$$

where

$$G_n = \frac{2n\pi}{a}, \qquad n = \pm1, \pm2, \pm3, \ldots \tag{10.11}$$

The V_n are the spatial Fourier coefficients. Integral multiples of a are the lattice vectors in the direct physical space. By analogy, integral multiples of $2\pi/a$, or G_n, are the lattice vectors in a "reciprocal lattice" k-space. This is much like expanding a time-varying electrical signal $\varepsilon(t + T) = \varepsilon(t)$ with a period T in a Fourier series:

$$\varepsilon(t) = \sum_{n = 0, \pm1, \pm2, \ldots} \varepsilon_n e^{in2\pi t/T}.$$

If the potential in the crystal is zero everywhere, or $V_{\mathrm{cr}}(x) = 0$, then the normalized solution (10.7a) of the Schrödinger equation (10.7) is simply:

$$\Psi^{(0)}_{E(k)}(x) = \sqrt{\frac{1}{L}}\, e^{ikx}, \qquad \text{where} \qquad E^{(0)}(k) = \frac{\hbar^2 k^2}{2m}. \tag{10.12}$$

The dispersion curve (or *E* vs. *k* curve) of the corresponding de Broglie wave is that of a free particle and is shown as the solid curve in Figure 10.3(a).

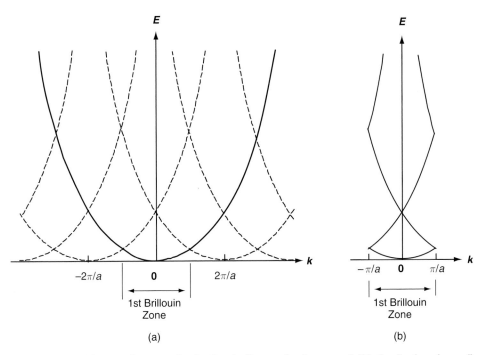

Figure 10.3 – (a) E vs. k curves in the "periodic-zone" scheme and (b) the "reduced-zone" scheme.

Introducing the periodic potential (10.10) as a perturbation, the corresponding eigen function and eigen value of the Schrödinger equation become, respectively, $\Psi_{E(k)}(x)$ and $E(k)$:

$$\left[-\frac{\hbar^2}{2m} \frac{\partial^2}{\partial x^2} + \sum_{n=\pm 1, \pm 2, \pm 3, \ldots} V_n e^{iG_n x} \right] \Psi_{E(k)}(x) = E(k)\, \Psi_{E(k)}(x). \tag{10.13}$$

This equation can be solved by the perturbation technique, as outlined in Section 9.1, if V_{cr} is a small perturbation, and the solution is of the form:

$$\Psi_{E(k)}(x) = \sqrt{\frac{1}{L}}\, e^{ikx} + \Psi^{(1)}_{E(k)}(x). \tag{10.14}$$

From (10.6a), according to the Bloch theorem, $\Psi^{(1)}_{E(k)}(x)$ must be of the form:

$$\Psi^{(1)}_{E(k)}(x) = \sum_{k'} C_{E(k)}(k') \sqrt{\frac{1}{L}}\, e^{ik'x}, \tag{10.15}$$

where

$$k' = G_n + k. \tag{10.15a}$$

Indeed, using the procedure of the time-independent perturbation theory for non-degenerate states, the first order perturbed solution is, from (9.6b) and in the limit of $L \to \infty$:

$$C_{E(k)}(k') = \frac{V_n}{E^{(0)}(k) - E^{(0)}(k')} \delta_{k',(k+G_n)}, \qquad (10.16)$$

for $E^{(0)}(k) \neq E^{(0)}(k')$ or $|k| \neq |k'|$ from (10.12); therefore, the wave function to the first order is:

$$\Psi_{E(k)}(x) = \sqrt{\frac{1}{L}} e^{ikx} + \Psi^{(1)}_{E(k)}(x) + \cdots$$

$$= \sqrt{\frac{1}{L}} e^{ikx} + \sum_{n=\pm 1, \pm 2, \pm 3, \ldots} \frac{V_n}{E^{(0)}(k) - E^{(0)}(k+G_n)} \sqrt{\frac{1}{L}} e^{i(k+G_n)x} + \cdots,$$

$$(10.17)$$

which is of the form (10.6a), as required by the Bloch theorem. From (9.7), there is no first order correction to the perturbed energy eigen values for a crystal potential of the form (10.10). (Note that there is no $n = 0$ term in the series expansion term of the crystal potential V_{cr} in (10.10).) The lowest order of correction is, therefore, the second order:

$$E(k) = E^{(0)}(k) + E^{(1)}(k) + E^{(2)}(k) + \cdots$$

$$= \frac{\hbar^2 k^2}{2m} + \sum_{n=1, 2, 3, \ldots} \frac{|V_n|^2}{E^{(0)}(k) - E^{(0)}(k+G_n)} + \cdots \qquad (10.18)$$

These results, (10.17) and (10.18), lead to two extremely important conclusions about the single-electron states in a periodic lattice:

1. In the periodic lattice, from (10.17), the eigen state corresponding to each energy value $E(k)$ is a Bloch state which is a sum of de Broglie waves of wave vectors $k + G_n$. The dispersion curves of the corresponding de Broglie waves are as shown in Figure 10.3(a). This is, of course, required by the Bloch theorem.
2. From (10.17) and (10.18), there are degeneracies in $E(k)$ at the k points where $E^{(0)}(k) = E^{(0)}(k+G_n)$ and $n = \pm 1, \pm 2, \pm 3, \ldots$ This occurs where the dispersion curves cross or, as shown, in Figure 10.3(a), where $k^2 = (k+G_n)^2$ or at the Brillouin zone boundaries $k = -G_n/2 = n\pi/a$ in k-space. Therefore, these results, (10.17) and (10.18), based on the non-degenerate perturbation theory, do not apply, and the degenerate perturbation theory must be used near the Brillouin zone boundaries. The regions around these crossing points are of critical importance for applications in semiconductor electronics and photonics; for this is where the band gap between the energy bands occurs.

The *E* vs. *k* curves shown in Figure 10.3(a) are unnecessarily repetitive. The same information can be gleaned from the restricted part of the curves within the first Brillouin zone between the boundaries at $\pm\pi/a$ showing multiple energy bands, as

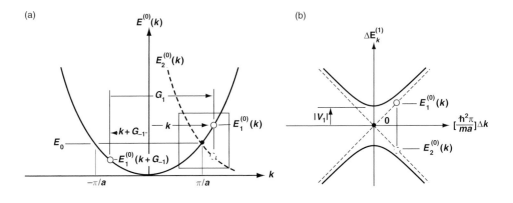

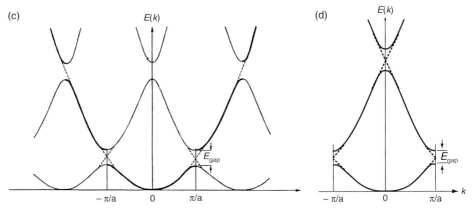

Figure 10.4 Schematics of E vs. K curves of (a) a free electron (heavy solid curve), and an electron in a periodic lattice (b)(solid curves) near the first Brillouin zone boundary (framed part of (a) with the inclusion of the first spatial Fourier harmonic $E_2^{(0)}(k)$), (c) shown in the "extended-zone scheme" [thick solid curves] and the "periodic-zone scheme" [thick plus thin solid curves], and (d) in the first Brillouin zone in the " reduced-zone scheme." See the discussion associated with Eq. (10.21) in the text.

shown in Figure 10.3(b). This presentation is called the "reduced-zone scheme," whereas the full diagram in Figure 10.3(a), is called the "periodic-zone scheme."

To use the degenerate perturbation theory developed in Section 9.1 to find the E vs. k curves near the crossing points, only a reduced set of basis states consisting of the unperturbed states of the same k-value but neighboring energy bands are needed. (For a non-perturbative alternative derivation of the following results, see, for example, Kittel (1996).) For example, let us designate $E^{(0)}(k) = \dfrac{\hbar^2 k^2}{2m}$, the solid curve in Figure 10.4(a), as $E_1^{(0)}(k)$ and its first spatial side-band curve shifted to the right of $E_1^{(0)}(k)$ by G_1 as $E_2^{(0)}(k) = \dfrac{(\hbar k - G_1)^2}{2m}$. $E_1^{(0)}(k)$ and $E_2^{(0)}(k)$ as functions of k are also shown as the dashed lines in Figure 10.4(b). Let the wave functions of the two states corresponding to $E_1^{(0)}(k)$ and $E_1^{(0)}(k - G_1)$ be $|1\rangle \equiv |\Psi_k^{(0)}\rangle$ and $|2\rangle \equiv |\Psi_{k-G_1}^{(0)}\rangle$. From the definitions of $E_1^{(0)}(k)$ and $E_2^{(0)}(k)$ given above, $E_1^{(0)}(k) = E_2^{(0)}(k + G_1)$. At the first Brillouin zone boundaries at $k = \pm\pi/a$, $E_1^{(0)}(k = \pi/a) = E_1^{(0)}(k = -\pi/a)$. Since $E_1^{(0)}(k = -\pi/a) = E_2^{(0)}(k = -\pi/a + G_1) = E_2^{(0)}(k = \pi/a)$, we have $E_1^{(0)}(k = \pi/a) = E_2^{(0)}(k = \pi/a)$ at the Brillouin-zone boundaries $k = \pm\pi/a$, as shown in Figure 10.4(a). *Near* the Brillouin zone boundary where $k = \frac{\pi}{a} + \Delta k$ and $\Delta k \ll k$, the eigen values are $E_1^{(0)}(k) \equiv E_0 + \Delta E_k^{(0)}$ and $E_2^{(0)}(k) \equiv E_0 - \Delta E_k^{(0)}$, where $\Delta E_k^{(0)} \cong \left(\frac{\hbar^2 \pi}{ma}\right)\Delta k$ and $E_0 \equiv E_1^{(0)}(\pi/a) = E_2^{(0)}(\pi/a)$, as shown in Figure 10.4(a) and 10.4(b). According to (9.11b), the corresponding 2×2 matrix to be diagonalized is:

$$\begin{pmatrix} E_0 + \Delta E_k^{(0)} & V_1 \\ V_1^* & E_0 - \Delta E_k^{(0)} \end{pmatrix} \begin{pmatrix} \langle \Psi_{E(k)}|1\rangle \\ \langle \Psi_{E(k)}|2\rangle \end{pmatrix}$$
$$= (E_0 + \Delta E_k^{(1)}) \begin{pmatrix} \langle \Psi_{E(k)}|1\rangle \\ \langle \Psi_{E(k)}|2\rangle \end{pmatrix} \tag{10.19}$$

where $|1\rangle \equiv |\Psi_{k=\pi/a}^{(0)}\rangle$ and $|2\rangle \equiv |\Psi_{k=(\pi/a)-a_1}^{(0)}\rangle$.

Solving (10.19) gives the characteristic equation for the two eigen values:

$$\left(\Delta E_k^{(1)}\right)^2 - \left(\Delta E_k^{(0)}\right)^2 = |V_1|^2. \tag{10.20}$$

Substituting in the zeroth-energy eigen value at the first Brillouin-zone boundary from (10.12) shows that the first order shift of the energy due to the periodic lattice as a function of k follows a hyperbolic equation with the unperturbed values as the asymptotes, as shown in Figure 10.4(b):

$$\left(\Delta E_k^{(1)}\right)^2 - \left(\frac{\hbar^4 \pi^2}{m^2 a^2}\right)\Delta k^2 = |V_1|^2, \tag{10.21}$$

and the two eigen values are:

$$\Delta E_{k\pm}^{(1)} = \pm\sqrt{\left(\frac{\hbar^4 \pi^2}{m^2 a^2}\right)\Delta k^2 + |V_1|^2} \tag{10.21a}$$

with an energy gap in-between. The new E vs. k curve in the "extended-zone scheme," the "periodic-zone scheme," and in the first Brillouin-zone of the "reduced-zone scheme" for the electron in the periodic lattice are all shown in Figure 10.4(c) and (d). The results in all three schemes contain basically the same information; therefore, it is usually sufficient just to show the results in one zone, using, for example, the reduced-zone scheme. The ratios of the expansion coefficients of corresponding wave functions are:

$$\frac{\langle \Psi_{E(k)\pm}|1\rangle}{\langle \Psi_{E(k)\pm}|2\rangle} = \frac{V_1}{\Delta E_k^{(1)} - \Delta E_{k\pm}^{(0)}}. \tag{10.22}$$

These results show that near the zone boundaries, there are energy gaps within which the electrons are "forbidden." Note that at the zone boundary, $\Delta E_k^{(0)} = 0$ and $\Delta E_{k\pm}^{(1)} = \pm|V_1|$. Since V_1 is purely real, the ratio of the coefficients in (10.22) is, therefore, either $+1$ or -1, or the perturbed wave functions are symmetric or anti-symmetric combinations of $|1\rangle = \sqrt{\dfrac{1}{L}}\, e^{i(\pi/a)x}$ and $|2\rangle = \sqrt{\dfrac{1}{L}}\, e^{-i(\pi/a)x}$. Thus, in the energy gaps at the Brillouin zone boundaries, the mixed waves are actually non-propagating standing waves. Physically, it is due to the strong back-scattering of the waves in the periodic lattice, when the lattice spacing is equal to integral multiples of the half-wavelength of the corresponding de Broglie waves of the electron at the zone boundaries.

We know that each momentum state, or k-state, has two spin states and can accommodate two electrons. Since the number of allowed momentum states or k-values within each Brillouin zone is exactly equal to the number of lattice points in the range L, if there is one atom per lattice point and one electron, or any odd number, of valence electrons per atom, half of the momentum states in the highest occupied energy band will be filled at zero degree temperature and the solid will be a metal. If each of the atoms has an even number of valence electrons, all the momentum states in the highest occupied energy band will be completely filled up to an energy band gap and the solid at zero degree will be an insulator or a semiconductor, depending upon the size of the energy band gap. The filled band is then the valence band and the next unfilled band is the conduction band.

The situation becomes considerably more complicated, however, when these considerations are extended to two or three dimensions. In two or three dimensions, some of the states of a given Brillouin zone in the k-space may actually have lower energies than some of the states in a lower Brillouin zone. It means that, in the reduced-zone scheme, the energy bands might overlap. In that case, these lower-energy states of the higher band might be filled before all the states in the lower band are filled. There is then no energy gap between the valence band and the conduction band and the crystal might then be a metal even if there are an even number of valence electrons per unit cell of the lattice. In real materials, the situation can be even more complicated. The E vs. k curves in different directions in three-dimensional space might be quite different. The bottom and the top of the bands may not occur at either the zone center or the zone boundaries, nor even at the same k-value. In the latter case, since radiative transitions between the valence and conduction band states tend to conserve the k-values according to the Fermi golden rule, this means that direct radiative recombination of an electron at the bottom of the conduction band and a hole at the top of the valence band are forbidden. In the context of optical applications, it implies, for example, that laser action in such solids as crystalline Si is unlikely to occur.

The electric conductivity of the solid depends on the transport properties of charge carriers in the solid. For metals, the outer electrons are freed from the individual atoms and can, therefore, roam freely in the entire solid and conduct electric currents at any temperature. For semiconductors at a finite temperature, the electrical characteristics of the material are primarily determined by the charge carriers in the states near the forbidden band-gap. This is because the electrons in the conduction band in thermal equilibrium at a finite temperature will settle near the bottom of the band and the holes left behind will rise to the top of the valence band. The charge-carriers will, thus, have many nearby vacant states in their respective bands to which they can move and become mobile in the solid. Thus, how the electrons are distributed over the available single-electron states of the lattice is essential to the understanding of the electronic properties of the solid.

10.4 Density-of-states and the Fermi energy for the free-electron gas model

In thermal equilibrium, the electrons in the solid will fill the available single-electron states of successively higher energy according to the principles of Fermi statistics, which is based on the energy of the electrons, as will be discussed in Section 10.5. The "density-of-states" gives the number of available states per differential energy-interval per unit volume of the solid as a function of either the energy or momentum of the single-electron states in the solid. For a solid of length L, the allowed momentum states subject to the cyclic boundary condition of Born and Von Karman are given by (10.9). Each value of N corresponds to an allowed momentum state, or the momentum states are separated by $2\pi/L$ in k-space. In the limit of large L, the corresponding values of $k = \pm N2\pi/L$ become a continuum. Thus, in one dimensional space, the total number of momentum states per unit physical length of the solid from $-k$ to $+k$ is simply $|k|/\pi$. Including the spin degeneracy of two, the total number of states per unit physical length in this range of k-space is $2|k|/\pi$. *For free particles* in the solid, $E = \hbar^2 k^2/2m$. Thus, the one-dimensional density-of-states including the two-fold spin-degeneracy is:

$$D^{(1)}(E) = \frac{\partial}{\partial E}\left(\frac{2|k|}{\pi}\right) = \frac{1}{\pi\hbar}\sqrt{\frac{2m}{E}}. \tag{10.23a}$$

The one-dimensional structure in the form of "quantum wires" is of considerable interest in modern electronics and photonics.

Generalizing to two dimensions, the corresponding density-of-states is:

$$D^{(2)}(E) = \frac{\partial}{\partial E}\left(\frac{|k|^2}{2\pi}\right) = \frac{m}{\pi\hbar^2}, \tag{10.23b}$$

which happens to be independent of E. This is a consequence of the fact that, for free particles, the energy E as well as the number of states per unit area in two-dimensional k-space are both proportional to k^2; thus, the number of states per area per energy interval is constant. This result is of importance in practical applications in modern electronic and photonic devices involving hetero-junctions and quantum wells that are modeled as two-dimensional electron gases.

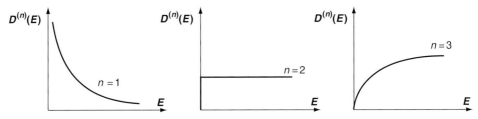

Figure 10.5 – Schematics of the densities-of-states for one-, two-, and three-dimensional free-electron gases.

For three dimensions, the corresponding density-of-states for *free particles* is:

$$D^{(3)}(E) = \frac{\partial}{\partial E}\left(\frac{|k|^3}{3\pi^2}\right) = \frac{\sqrt{2m^3}}{\pi^2 \hbar^3}\sqrt{E}. \tag{10.23c}$$

It is applicable to metals. For comparison, the densities-of-states for the free-electron gas (10.23a, b, c) are shown qualitatively in Figure 10.5. In a periodic lattice, however, because the E vs. k curve is different from that of free particles, the density-of-states as a function of E must be modified accordingly, as will be discussed later.

The highest energy level occupied by the valence electrons in a solid at 0 K is called the "Fermi energy," E_F. Using again the one-dimensional free-particle model, the Fermi energy can be determined from the number of valence electrons, N_e, per unit length of the solid:

$$N_e = \int_0^{E_F} D^{(1)}(E)\mathrm{d}E = \frac{2\sqrt{2m_e}}{\pi\hbar}\sqrt{E_F} \tag{10.24}$$

from (10.23a), and:

$$E_F = \frac{\hbar^2}{2m_e}\left(\frac{\pi}{2}N_e\right)^2, \tag{10.25}$$

where m_e is the electron mass. The corresponding Fermi energy in terms of the number of electrons per volume, N_e, for a three-dimensional solid is:

$$E_F = \frac{\hbar^2}{2m_e}(3\pi^2 N_e)^{3/2}, \tag{10.26}$$

from (10.23c). (For the two dimensional case, see Problem 10.1). Thus, the Fermi energy is known from the number density of the atoms in the solid and the number of valence electrons per atom. Knowing the Fermi energy is tantamount to knowing the valence-electron density, and vice versa.

10.5   Fermi–Dirac Distribution function and the chemical potential

In the limit of 0 K temperature, all the electrons in the solid will occupy the lowest possible energy state, subject to the Pauli exclusion principle. The corresponding

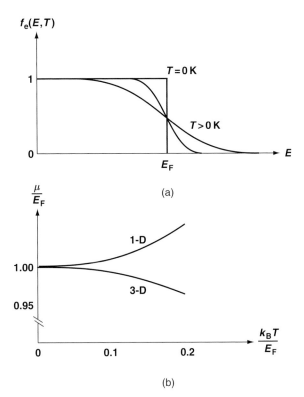

Figure 10.6 (a) Fermi–Dirac distribution functions at $T = 0$ K and at two $T > 0$ K. (b) Dependence of the chemical potential on the temperature T for one-dimensional (1-D) and three-dimensional (3-D) systems.

distribution function as a function of energy is shown in Figure 10.6(a). It is equal to one up to the Fermi energy and drops to zero above the Fermi energy.

At a finite temperature, some of the electrons will be excited to states above the Fermi energy. The probability that a given energy state is occupied by fermions follows the "Fermi–Dirac distribution function":

$$f_e(E, T) = \frac{1}{e^{(E-\mu)/k_B T} + 1},$$

(10.27)

where K_B is the Boltzmann constant and is equal to 1.38×10^{-16} erg / K. μ is the "chemical potential," which is by definition the value of E at which the probability of occupation is equal to one-half, or $f_e(\mu, T) \equiv 1/2$. Examples of the Fermi–Dirac distribution function for the electrons are shown in Figure 10.6(a). The probability distribution, $f_h(E, T)$, of the holes left behind below the Fermi level by the thermally excited electrons is:

$$f_h(E, T) = 1 - f_e(E, T) = 1 - \frac{1}{e^{(E-\mu)/k_B T} + 1}$$

$$= \frac{1}{e^{(\mu-E)/k_B T} + 1}.$$

(10.28)

For $T > 0$, the Fermi–Dirac distribution functions for the holes and electrons are approximately symmetric with respect to the chemical potential, as can also be seen qualitatively in Figure 10.6(a). Physically, the chemical potential is to mass flow as the electrical potential is to electrical current flow. According to (10.27), the electron concentration at a given electron energy level above the chemical potential is higher where the chemical potential is higher; electrons will, therefore, diffuse spatially from where the chemical potential is higher to where it is lower.

In the limit of $T = 0$ K, the Fermi–Dirac distribution function is discontinuous and the chemical potential is by definition equal to the Fermi energy. At a finite temperature, the chemical potential is determined from the condition that the total number of valence electrons N_e remains the same as the temperature changes:

$$N_e = \int_0^{E_F} D(E)\mathrm{d}E = \int_0^{\infty} D(E)f_e(E, T)\mathrm{d}E = \int_0^{\infty} \frac{D(E)}{e^{(E-\mu)/k_B T} + 1} \, \mathrm{d}E, \tag{10.29}$$

with the bottom of the valence band chosen as $E = 0$. The chemical potential as a function of the temperature T can then be determined by solving (10.29) by using the appropriate density-of-states $D(E)$. For example, the results so obtained for one- and three-dimensional chemical potentials are shown qualitatively in Figure 10.6(b) for free-electron gas (see Problem 10.2 for the corresponding two-dimensional result). As can be seen, it is approximately equal to the Fermi energy for most of the temperature range of practical interest. For most applications in semiconductor electronics and photonics there is often, therefore, no need to make a distinction between the chemical potential and the Fermi energy. In such cases, the Fermi–Dirac distributions are simply given as:

$$f_e(E, T) \cong \frac{1}{e^{(E-E_F)/k_B T} + 1}, \tag{10.27a}$$

for the electrons, and

$$f_h(E, T) \cong \frac{1}{e^{(E_F-E)/k_B T} + 1}, \tag{10.28a}$$

for the holes, respectively. E_F in (10.27a) and (10.28a) is commonly referred as the "Fermi level," which is often used to characterize the spatial variations of the carrier concentration in, for example, p–n junctions, as will be discussed in detail later in this chapter.

As long as the density-of-states is non-zero and a continuous function of E, as in the case of a free-electron gas, this definition of the chemical potential is unambiguous and its location can be determined by solving Eq. (10.29). For semiconductors, however, there is an energy gap between the valence and conduction bands. The location of the chemical potential in the gap cannot be determined on the basis of (10.29) when the temperature is at exactly $T = 0$ K. This is because, at $T = 0$ K exactly, the valence band is fully occupied and the conduction band is completely empty. All one knows is that the chemical potential must be somewhere in the energy gap between the conduction band and the valence band. Exactly where it is cannot be determined from Eq. (10.29), because it is an identity independent of μ at exactly $T = 0$ K. On the other hand, when the temperature is even infinitesimally above 0 k, (10.29) becomes:

$$\int_0^{E_v} D(E)dE = \int_0^{E_v} \frac{D(E)}{e^{(E-\mu)/k_BT} + 1} \, dE + \int_{E_c}^{\infty} \frac{D(E)}{e^{(E-\mu)/k_BT} + 1} \, dE,$$

which leads to:

$$\int_0^{E_v} D(E)\left[1 - \frac{1}{e^{(E-\mu)/k_BT} + 1}\right]dE = \int_0^{E_v} D(E)\left[\frac{1}{e^{(\mu-E)/k_BT} + 1}\right]dE$$

$$= \int_{E_c}^{\infty} \frac{D(E)}{e^{(E-\mu)/k_BT} + 1} \, dE.$$

Making use of (10.27) and (10.28), the above equation becomes:

$$\int_0^{E_v} D(E)f_h(E, T)dE = \int_{E_c}^{\infty} D(E)f_e(E, T)dE. \tag{10.30}$$

The left hand side of (10.30) gives the number of holes in the valence band, and the right hand side of (10.30) gives the number of electrons in the conduction band. Given in this form, it can be seen from Figure 10.3(a) that as long as the width of the valence band is much larger than K_BT, which is usually the case in the practical situations of interest, the lower limit on the left side of (10.30) can then be taken to be $-\infty$ and the choice of the reference energy $E = 0$ can be arbitrary:

$$\int_{-\infty}^{E_v} D_v(E)f_h(E)dE = \int_{E_c}^{\infty} D_c(E)f_e(E)dE, \tag{10.30a}$$

which says that the total number of electrons in the conduction band is equal to the total number of holes in the valence band. This equation gives a very important condition, which is known as the "charge-neutrality condition." It reflects the obvious fact that each electron that gets excited into the conduction band leaves a hole in the valence band, and the semiconductor remains electrically neutral.

The charge neutrality condition, (10.30a), is the basic condition that determines the location of the chemical potential in semiconductors in general. To do so, it is necessary to know $D_c(E)$ and $D_v(E)$ as functions of E over their respective range of integration, where $f_e(E)$ and $f_h(E)$ are appreciable. For semiconductors, it is just above and below the band gap, or near the bottom of the conduction band for the electrons and the top of the valence band for holes, respectively. In the special case where the curvatures of their respective E vs. k curves are the same, as in Figure 10.4(b), for, example, and at $T = 0^+$ K, it follows from the charge-neutrality condition (10.30) that:

$$\mu = \frac{E_c + E_v}{2}. \tag{10.31}$$

For temperatures $T > 0$ K, the chemical potential for semiconductors in general will change with T and depend on the band structures of the conduction and valence bands, as will be discussed in more detail in Section 10.6.

Note that, when $|E - \mu| \gg 3k_B T$, the Fermi–Dirac distribution functions (10.27) and (10.28) can be approximated, respectively, by the classical Boltzmann distribution functions:

$$f_e(E, T) \approx e^{-(E - E_F)/k_B T}, \tag{10.27b}$$

for the conduction-band electrons and

$$f_h(E, T) \approx e^{-(E_F - E)/k_B T}, \tag{10.28b}$$

for the valence-band holes. Semiconductors under these conditions are described qualitatively as being "non-degenerate." It is necessary to use the more exact Fermi distribution when the semiconductor is under "degenerate" conditions.

With the Fermi–Dirac distributions of the electrons and holes and their respective densities-of-states, it is possible to determine the density of the electrons, n_c, in the conduction band and the density of the holes, p_v, in the valence band at any spatial point in the semiconductor once the densities-of-states and the Fermi level at that point are known:

$$n_c = \int_{E_c}^{\infty} D_c(E) f_e(E) dE, \tag{10.32a}$$

and

$$p_v = \int_{-\infty}^{E_v} D_v(E) f_h(E) dE. \tag{10.32b}$$

When $(E - E_F) \gg k_B T$, the electron concentration (10.32a) can be written as:

$$
\begin{aligned}
n_c &= \int_{E_c}^{\infty} D_c(E) f_e(E) dE \\
&\cong e^{-(E_c - E_F)/k_B T} \int_{E_c}^{\infty} e^{-(E - E_c)/k_B T} D_c(E) dE, \\
&\equiv N_c e^{-(E_c - E_F)/k_B T}
\end{aligned}
\tag{10.33}
$$

where

$$N_c = \int_{E_c}^{\infty} e^{-(E - E_c)/k_B T} D_c(E) dE. \tag{10.33a}$$

Equation (10.33) shows that the electrons in the conduction band can be equivalently viewed as all concentrated at the bottom of the conduction band where the equivalent density-of-states is the "effective density-of-states," N_c, which is a number independent of E. A similar effective density-of-states can be defined for the holes:

$$p_v = \int_{-\infty}^{E_v} D_v(E)f_h(E)\mathrm{d}E$$

$$\cong e^{-(E_F-E_v)/k_BT} \int_{-\infty}^{E_v} e^{-(E_v-E)/k_BT} D_v(E)\mathrm{d}E$$

$$\equiv N_v e^{-(E_F-E_v)/k_BT}, \qquad (10.34)$$

where

$$N_v = \int_{-\infty}^{E_v} e^{-(E_v-E)/k_BT} D_v(E)\mathrm{d}E. \qquad (10.34a)$$

Note that both the carrier concentrations and the effective densities-of-states as given in the expressions (10.33) to (10.34a) are independent of the choice of where the reference $E = 0$ is, and the integrals are defined relative to the bottom of the conduction band E_c or the top of the valence band E_v. From (10.33) and (10.34), it can be seen that the product of the electron concentration in the conduction band and the hole concentration in the valence band is independent of the position of the Fermi level:

$$n_c p_v = N_c N_v e^{-E_g/k_BT}, \qquad (10.35)$$

and depends only on the band-gap E_g. In the case of intrinsic semiconductors, the electron and hole densities are the same, or $n_c = p_v \equiv n_i$; therefore:

$$n_i^2 = N_c N_v e^{-E_g/k_BT}. \qquad (10.36)$$

The right side of Eq. (10.35) is independent of the sources of the charge carriers in the conduction band and the valence band; it applies, therefore, to intrinsic as well as extrinsic semiconductors (extrinsic semiconductors are intrinsic semiconductors doped with impurity atoms, as will be discussed in Section 10.7). For extrinsic semiconductors, where n_c & $p_v \neq n_i$, it follows from (10.35) and (10.36) that:

$$n_i^2 = n_c p_v. \qquad (10.37)$$

Equation (10.37) is a very useful practical result. It shows that the product of the electrons in the conduction band and the holes in the valence band is a constant independent of the sources of the charges or the location of the Fermi level and, hence, the relative concentrations of the electrons and holes in the semiconductor. Therefore, increasing the concentration of one type of carriers by suitable doping of impurity atoms will necessarily lead to a proportional decrease in the concentration of the oppositely charged carriers.

Finally, the Fermi level in intrinsic semiconductors can be determined from the condition $n_c = p_v$, which is the "charge neutrality condition":

$$\int_{-\infty}^{E_v} D_v(E)f_h(E)\mathrm{d}E = \int_{E_c}^{\infty} D_c(E)f_e(E)\mathrm{d}E, \qquad (10.30a)$$

from (10.33) and (10.34). In the limit of $|E - E_F \gg k_B T$ and making use of (10.33a) and (10.34a), the charge neutrality condition gives:

$$E_F = \frac{E_c + E_v}{2} + \frac{1}{2} k_B T \ln\left(\frac{N_v}{N_c}\right). \tag{10.38}$$

To determine where the Fermi level is, we must now determine the effective densities-of-states in terms of the band structures of the semiconductor. For that, the concept of effective mass of electrons and holes in semiconductors is useful and will be introduced next.

10.6 Effective mass of electrons and holes and group velocity in semiconductors

At the bottom of the conduction band, the E vs. k curve is parabolic, as shown in Figure 10.4(b)–(d), which can be put in the form:

$$E_c(k) = \frac{\hbar^2}{2m_e^*} k^2, \tag{10.39}$$

where m_e^* is a measure of the curvature of the E vs. k curve:

$$m_e^* = \hbar^2 \left(\frac{\partial^2 E_c}{\partial k^2}\right)^{-1}, \tag{10.40}$$

and plays the role of the particle mass in the E vs. k curve for the free electron as in (10.12). It can, therefore, be thought of as the effective mass of the electron in the periodic lattice. That it can indeed be so interpreted can be seen from another point of view.

As discussed in Chapter 3, a spatially localized electron is a wave packet. The group velocity v_g of the corresponding wave packet is:

$$v_g = \frac{\partial \omega}{\partial k} = \frac{1}{\hbar} \frac{\partial E_c}{\partial k}. \tag{10.41}$$

The equation of motion of a localized electron near the bottom of the conduction band subject to a force F is:

$$F = \frac{dp}{dt}, \tag{10.42}$$

where p is the momentum of the particle, which for de Broglie waves is:

$$F = \frac{dp}{dt} = \hbar \frac{dk}{dt}. \tag{10.43}$$

For a wave packet to be viewed as a particle, the momentum p can also be defined as the product of an effective mass m_e^* and the group velocity v_g:

$$F = m_e^* \frac{dv_g}{dt} = \frac{m_e^*}{\hbar} \left(\frac{\partial^2 E_c}{\partial k^2} \right) \frac{dk}{dt}, \tag{10.44}$$

from (10.41) and (10.42). For (10.43) and (10.44) to be consistent, we must have:

$$\frac{\partial^2 E_c}{\partial k^2} = \frac{\hbar^2}{m_e^*} \qquad \text{or} \qquad m_e^* = \hbar^2 \left(\frac{\partial^2 E_c}{\partial k^2} \right)^{-1}, \tag{10.40a}$$

as in (10.40). Similarly, the effective mass of the hole is:

$$m_h^* = \hbar^2 \left(\frac{\partial^2 E_v}{\partial k^2} \right)^{-1}. \tag{10.45}$$

Note that the energy of the hole E_v at a given k-value is equal to the negative of the corresponding energy E_e of the missing electron from the valence band at the same k-value, or $E_v(k) = -E_e(k)$.

With the effective mass, it is now also possible to give an explicit expression of the effective density-of-states in terms of the effective mass of the electrons at the bottom of the conduction band, from (10.23c) and (10.33a):

$$N_c = \frac{1}{\sqrt{2}} \left(\frac{m_e^* k_B T}{\pi^2 \hbar^2} \right)^{3/2}, \tag{10.46}$$

by replacing the free-electron mass m by the effective mass of the electron m_e^* in (10.23c). Similarly, the effective density-of-states of the holes at the top of the valence band is:

$$N_v = \frac{1}{\sqrt{2}} \left(\frac{m_h^* k_B T}{\pi^2 \hbar^2} \right)^{3/2}. \tag{10.47}$$

The position of the Fermi level can then be determined from effective masses of the electrons and holes on the basis of Eqs. (10.38), (10.46) and (10.47):

$$E_F \cong \frac{E_c + E_v}{2} + \frac{3kT}{4} \ln \left(\frac{m_h^*}{m_e^*} \right). \tag{10.48}$$

Thus, if the effective mass of the electron at the bottom of the conduction band and that of the hole at the top of the valence band are the same, the Fermi level must be at the middle of the energy gap separating the conduction band and the valence bands, as shown in Figure 10.7(a), because the number of electrons is equal to the number of holes. If the effective mass of the electrons m_e^* near the bottom of the conduction band is smaller than that of the holes m_h^* at the top of the valence band, then the Fermi level must be above the middle; this is the case of GaAs, as shown in Figure 10.7(c). Otherwise, the Fermi level is at or below the middle.

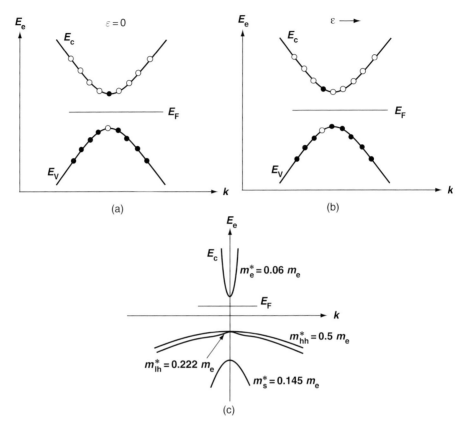

Figure 10.7 (a) Schematic of the electron and hole states and the E_F level for a semiconductor with equal electron and hole effective masses. (b) Schematic showing the effect of an applied electric field $\vec{\varepsilon}$ on the position of the occupied conduction band electron state and the unoccupied electron state in the valence band; both move in the direction opposite to that of $\vec{\varepsilon}$. (Open circle: unfilled electron state; solid circle: occupied electron state.) (c) Schematic of the band structure of GaAs consisting of the conduction band, the heavy-hole (hh), light-hole (lh), and split-off (s) valence bands. The location of the Fermi-level depends on the temperature according to the charge-neutrality condition.

It is also of interest to consider the directions of the motions of the electrons and holes under the influence of an applied electric field $\vec{\varepsilon}$ in, for example, the $+x$ direction. For the electron, the corresponding force is equal to $\vec{F}_e = -e\,\vec{\varepsilon}$ and is in the $-x$ direction. The change in the momentum state of the electron $\vec{k}_e$ follows from the

equation: $\dfrac{\mathrm{d}\hbar\vec{k}_e}{\mathrm{d}t} = \vec{F}_e$; it is in the $-k_x$ direction in k-space. At the bottom of a parabolic conduction band, the effective mass of the electron is always positive; therefore, from the equation of motion (10.44), the rate of change of the group velocity must be negative, or $\mathrm{d}\vec{v}_g/\mathrm{d}t < 0$. Also, at the bottom of the conduction band, as the momentum $\hbar\vec{k}_e$ becomes negative, $\partial E_e/\partial k_e$ becomes more negative; therefore, the group velocity $\vec{v}_g$ is in the $-x$ direction. The case for the hole is tricky. First, although the force $\vec{F}_h = +e\vec{\varepsilon}$ is now in the $+x$ direction, the position of the hole in the valence band in

Table 10.1. *Relative directions of the motions of the electrons at the bottom of a parabolic conduction band and holes at the top of a parabolic valence band induced by an applied electric field in the +x direction.*

	Electron	Hole
$\vec{\varepsilon}$	$\rightarrow$	$\rightarrow$
$\vec{F}$	$\leftarrow$	$\rightarrow$
$\hbar\vec{k}$	$\leftarrow$	$\rightarrow$
m^*	>0	>0
$\mathrm{d}\vec{v}_g/\mathrm{d}t$	<0	>0
$\vec{v}_g$	$\leftarrow$	$\rightarrow$
$\vec{J}$	$\rightarrow$	$\rightarrow$

k-space moves in the $-k_x$ direction. The reason for this is that all the electrons move in lock-steps in the $-k_x$ direction; thus, the state of the momentum of the missing electron $\vec{k}_e$, which is where the hole is in k-space, also moves in the $-k_x$ direction (see Figure 10.7(b)). For a completely filled valence band, application of an electric field does not change the total momentum of all the electrons in the valence band. Missing an electron with a negative momentum from the valence band means, therefore, the subtraction of an electron with a negative momentum from the filled valence band, which means the sum total of the momentum of the remaining electrons in the valence band, or that of the hole, must be positive or in the $+k_h$ direction, as shown in Table 10.1. The group velocity of the hole, $\vec{v}_g = \dfrac{\partial E_h}{\partial k_h} = \dfrac{\partial E_e}{\partial k_e}$ at the top of valence band is in the $+x$ direction. Because the force $\vec{F}_h$ acting on the hole as well as the effective mass of the hole are positive, $\mathrm{d}\vec{v}_g/\mathrm{d}t > 0$ as required by the equation of motion $\vec{F}_h = m_h^* \mathrm{d}\vec{v}_g/\mathrm{d}t$. Here, we must be careful about the direction of the movement of the wave packet, that is the hole in physical space (x), which is determined by the direction of the group velocity, and the direction of the momentum of the hole in k-space, which is determined by the direction of $\vec{k}_h$, or the phase velocity of the corresponding de Broglie wave. These conclusions are summarized in Table 10.1. An important consequence of all of these considerations is that the corresponding electric current $\vec{J}$ is always in the $+x$ direction for both electron and hole currents, as is to be expected.

All the results obtained so far are primarily for intrinsic semiconductors. The electronic properties of the material can, however, be drastically altered by the presence of impurities. For this, we need to consider extrinsic semiconductors containing n-type or p-type dopants.

10.7 n-type and p-type extrinsic semiconductors

When some of the group IV atoms in a IV–IV "intrinsic" semiconductor crystal, such as Si or Ge, are substituted by group V or III atoms, such as As or Ga, these impurity atoms can act as "donors" or "acceptors," respectively, of electrons in the "extrinsic" semiconductor.

If a group IV atom is substituted by a group V atom with five valence electrons, the s and p valence-orbitals of the V atom will form four sp^3 orbitals of a tetrahedral complex and bond with the group IV atoms as a part of the diamond lattice of the extrinsic crystal. Four of the five valence electrons of the group V atom will occupy these valence-bond states; the one remaining valence electron will be loosely bound to the positively charged impurity V-ion as a hydrogenic atom with a relatively small ionization energy compared to the band gap of the semiconductor. The potential due to the positively charged nucleus of this impurity ion in the host lattice experienced by this loosely attached electron in the hydrogenic model of this dopant is approximately a Coulomb potential in a dielectric medium; it is, therefore:

$$V_i(r) \approx -\frac{e^2}{\epsilon r},$$ (10.49)

where ϵ is the dielectric constant of the host crystal. The corresponding energy levels of the hydrogenic ion are then, from (6.37):

$$E_n \cong -\frac{e^4 m_e^*}{2\epsilon^2 \hbar^2 n^2},$$ (10.50)

where n is the corresponding principal quantum number. The numerical value of ϵ for a typical semiconductor is on the order of 10. In addition, the effective mass m_e^* can also often be considerably smaller than the free-electron mass. Thus, the ionization energy of such an impurity level can be much less than the energy gap between the conduction and valence bands of the intrinsic semiconductor, and is typically on the order of tens of meV. Also, the corresponding Bohr orbit of the hydrogenic model of the dopant in the crystal will become much larger than that of the hydrogen atom in free space. It depends, of course, on the nature of the impurity atom and the host semiconductor. There can also be "deep donors or acceptors" that are located closer to the middle of the band gap. Numerically, in the case of arsenic atom in Si, for example, it is ~50 meV, as compared to a band gap of 1.1 eV for Si. This impurity atom can, thus, be easily ionized and donate its fifth valence electron to the conduction band of the extrinsic semiconductor, leaving an As$^+$ ion in place of a Si atom at one of the atomic sites in the diamond lattice. The crystal will then have more mobile negatively charged electrons in the conduction band to conduct electricity than the intrinsic Si and is, therefore, of the "n-type extrinsic semiconductor." Thus, As atoms act like donors in Si or Ge crystals, and Si or Ge crystals doped with As impurity atoms are "n-type" semiconductors. Similarly, if a group IV atom in a IV–IV semiconductor is substituted by a group III atom, a hole (or a deficit of one electron) is present and can accept an electron from somewhere. For example, the group III Ga atom is an acceptor in a Si or Ge crystal. The "ionization" energy of Ga in Si is ~65 meV. Si or Ge doped with Ga is then a "p-type" semiconductor.

Semiconductors doped with donor or acceptor atoms are still electrically neutral – the charge-neutrality condition still holds for extrinsic semiconductors. Thus, in

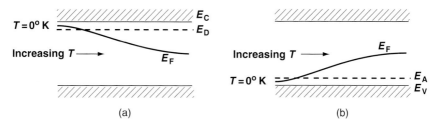

Figure 10.8 Variation of the Fermi level with temperature for: (a) n-type semiconductors ($N_d \neq 0$) and (b) p-type semiconductors ($N_a \neq 0$).

n-type semiconductors, the total number of negative charges in the conduction band n_c must be equal to the total number of holes p_v in the valence band plus the spatially fixed ionized donors N_d^+ in the n-type semiconductor of donor concentration N_d:

$$n_c = p_v + N_d^+. \tag{10.51}$$

Similarly, for a p-type semiconductor of acceptor concentration N_a, the charge-neutrality condition is:

$$p_v = n_c + N_a^-. \tag{10.52}$$

By appropriately including the donor and acceptor states in the density-of-states in (10.30a), it is possible to calculate the location of the Fermi levels for n-type or p-type semiconductors once the positions of the donor and acceptor levels in the band gap are known. The detailed numerical procedure for such calculations is algebraically involved and not particularly instructive for the present purpose. Qualitatively, based on (10.37), increasing the donor concentration must mean a decrease in the hole concentration and, therefore, the corresponding Fermi level must move up. Similarly, increasing the acceptor concentration means that the Fermi level must move down.

The position of the Fermi level as a function of temperature in the n-type crystal is shown qualitatively in Figure 10.8(a). At low temperatures, such that the thermal energy $k_B T$ is small compared to the band gap but comparable to the ionization energy of the donor, one expects the conduction band electrons to be mostly from the donors. Thus, in the n-type materials, the Fermi level at low temperatures will be close to the middle between the donor level and the bottom of the conduction band. As the temperature increases, more and more of the conduction band electrons will come from the valence band and the Fermi level of the extrinsic material will move downward and eventually approach that of the intrinsic material. Similarly, for the p-type material, the Fermi level at low temperatures will be close to the middle between the acceptor level and the top of the valence band and will rise to that of the intrinsic material as the temperature increases. When the semiconductor is heavily doped, the Fermi level could be within, for example, $\sim 3 k_B T$ of the bottom of the conduction band or the top of the valence band. Furthermore, the presence of a large number of impurities in the semiconductor could significantly modify the band structures of the crystal. Band "tails" at the bottom of the conduction band and the top of the valence band could appear as a result. It is then said to be "degenerately-doped." In this case, the carrier distributions in the conduction or valence band are described by the more exact Fermi–Dirac distribution functions, not the approximate Maxwell–Boltzmann distribution functions.

10.8 The p–n junction

When the p-type and n-type doped semiconductors of the same kind (Figure 10.9(a)) are brought into contact to form a p–n junction (Figure 10.9(b)), because of the difference in the Fermi levels, or the chemical potentials, charge-carriers will flow from one side to the other, leaving spatially fixed ionized donors and acceptors behind, which results in a built-in electrical potential difference across the junction. Such a p–n junction is of great technological importance.

The Fermi–Dirac distribution functions, (10.27) and (10.28), show that the carrier concentrations in the semiconductor are determined by the local chemical potential. At a p–n junction, before there is any transfer of charges from one side to the other, both sides are electrically neutral. The states of the same energy relative to the top of the valence band on the two sides of the junction can, however, have very different electron or hole populations because of the difference in the chemical potentials (Figure 10.9(a)). When the two sides are brought together and form a 'perfect' junction, as a result of the concentration gradients of the conduction band electrons and the valence band holes, charge-carriers will move across the junction. Conduction band electrons will diffuse across the junction from the n-side to the p-side to be trapped by the acceptors on the p-side and leave positively charged ionized donors on the n-side. Similarly, valence band holes will also diffuse from the p-side of the

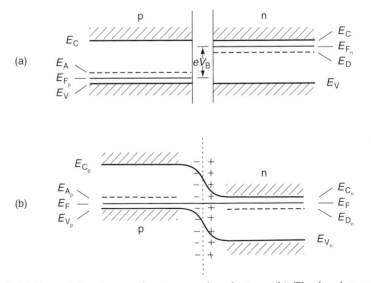

Figure 10.9 (a) Separated p-type and n-type semiconductors. (b) The band structure and spatially fixed space-charge (indicated by + and – signs) in the p–n junction region.

junction to the n-side to be captured by the donors on the n-side and leave negatively charged acceptors on the p-side. The resulting space-charge field due to the spatially fixed ionized donors on the n-side and the ionized acceptors on the p-side will raise the electron energy on the p-side relative to that of the n-side until the chemical potentials on the two sides become equalized. Under such a condition, the probabilities of occupation of the states of the same energy relative to the common chemical potential on the two sides are equal and no charge will flow across the junction. As a result, the electron energy corresponding to the bottom of the conduction band on the p-side, E_{c_p}, is higher than that on the n-side, as shown in Figure 10.9(b). The difference is the "built-in electron potential":

$$eV_B = E_{c_p} - E_{c_n}. \tag{10.53}$$

Note that V_B in (10.53) is defined as an electrical potential (not potential energy as elsewhere earlier in the book) and has a positive value, since E_c refers to the *electron energy* at the bottom of the conduction band.

Doping gradients can also lead to built-in fields for the minority carriers (for example, in the base region in transistors). Gradients in the composition of semiconductor lead to a band gap that changes with position. This also gives strong built-in fields sometimes. These tricks are used in practical semiconductor devices.

The p–n junction has an asymmetric voltage-current characteristic (Figure 10.10(a)), in that the electric current-flow across the junction from the p-side to the n-side is very much larger when the applied voltage on the p-side is positive relative to that on the n-side (see Figure 10.10(b)) than the reverse current from the n-side to the p-side when the junction is reverse-biased. This asymmetric nature of the p–n junction

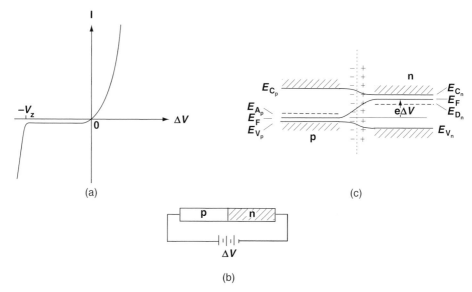

Figure 10.10 (a) I–V characteristics of a p–n junction. (b) Forward biased p–n junction (c) The corresponding band diagram.

can be understood by referring first to Figure 10.9. In the absence of the applied voltage, the electron concentration on the n-side (majority carrier) of the junction is, within the Boltzmann approximation:

$$n_n \cong N_c e^{-(E_{cn}-E_F)/k_B T},\tag{10.54}$$

and the electron concentration on the p-side (minority carrier) is:

$$n_p \cong N_c e^{-(E_{cp}-E_F)/k_B T}.\tag{10.55}$$

The ratio of the electron densities on the two sides is, therefore:

$$\frac{n_p}{n_n} \cong e^{-(E_{cp}-E_{cn})/k_B T} = e^{-eV_B/k_B T},\tag{10.56}$$

which is usually a number much less than 1.

Suppose now the electrical potential on the p-side is raised by an applied voltage source by ΔV, or the *electron energy* on the p-side is *lowered* by ($e\Delta V$) relative to the n-side, as shown in Figure (10.10(c)). With the applied voltage, the carriers are now in a non-equilibrium situation in the junction region where the Fermi level concept no longer applies. The electron and hole densities are now determined locally by their separate "quasi-Fermi levels," which depend on the applied electric potential and the local electric field due to the space-charge effect of the ionized donors and acceptors in the junction region. Apart from the details of the spatial variations of the non-equilibrium charges, the excess electron density Δn_c in the conduction band on the n-side above that of the corresponding state on the p-side of the junction region is, from (10.56):

$$\Delta n_c = n_p(e^{e\Delta V/k_B T} - 1).\tag{10.57}$$

Because the corresponding change in the potential in the junction region and the concomitant excess electrons on the n-side are maintained by an external voltage source, electrons must be constantly supplied to the n-side and drained from the p-side. There is, therefore, a constant (electric) diffusion current (Note this "diffusion current" is not to be confused with the "ballistic current" defined in (4.9)) through the junction from the p-side to the n-side. From the solution of the appropriate diffusion equation and (10.57), this diffusion current due to the electrons is:

$$J_e = \frac{en_p D_e}{L_n}(e^{e\Delta V/k_B T} - 1),\tag{10.58}$$

where D_e and L_n are the diffusion coefficient and diffusion length of the electrons, respectively. (See, for example, Smith (1964), p. 270.) Similarly, there is an excess hole concentration on the p-side of the junction equal to:

$$\Delta p_v = p_n(e^{e\Delta V/k_B T} - 1),\tag{10.59}$$

and a diffusion current due to the holes through the junction equal to:

$$J_h = \frac{ep_n D_h}{L_p}(e^{e\Delta V/k_B T} - 1),\tag{10.60}$$

where D_h and L_p are the diffusion coefficient and length of the holes, respectively. If the space-charge region is sufficiently narrow so that the recombination of the electrons and the holes in it has negligible effect on the currents in either regions, we can assume J_e and J_h to be continuous across the junction, and the total electric current J flowing through the external voltage source is, thus:

$$J = J_s \left[e^{e\Delta V/k_B T} - 1 \right], \tag{10.61}$$

where

$$J_s \equiv \frac{e n_p D_e}{L_n} + \frac{e p_n D_h}{L_p}, \tag{10.62}$$

is a "saturation current." This equation is frequently referred to as the Shockley equation. It shows clearly an asymmetric current characteristic: if ΔV is positive, or the junction is forward-biased, the current in the forward direction (electrons from the n-side to the p-side) increases exponentially. If the junction is reverse-biased (or ΔV is negative) the reverse current is limited by the "saturation current" J_s, until ΔV reaches a break-down voltage V_z due to avalanche multiplication.

Qualitatively, as can be seen from Figure 10.10(b), increasing the forward-bias voltage reduces the electron energy difference between the bottom of the conduction band on the p-side and that on the n-side. This allows more of the large number of electrons closer to the conduction band edge on the n-side to diffuse, or be "injected," into the p-side, and similarly for the large number of holes near the valence band edge on the p-side to be injected into the n-side. On the other hand, when the junction is reverse-biased, the electron potential difference between the two sides becomes even larger. The number of electrons in the conduction band on the p-side far above the Fermi level now becomes greater than the corresponding electron population of the same potential on the n-side and will diffuse into the n-side. However, because the number of electrons on the p-side without the reverse-bias is small to begin with, the reverse current is much smaller than the forward current in the forward-biased case. Similarly, the much smaller number of holes on the n-side far below the Fermi level will diffuse toward the p-side in the direction according to the externally applied electric field. This also contributes to the much smaller reverse current, as shown in (10.62). Hetero-junctions with materials of different band-gaps and work functions can lead to junctions with virtually zero back current.

The above description of what happens at the p–n junction is a highly simplistic picture. The junctions in practical devices are much more complicated, due to absorbed gas atoms from the ambient giving electronic surface states which bend the bands. They can lead to inversion layers bypassing a junction. Oxygen and water are two different absorbants which act as donors and acceptors, respectively. Therefore, "passivation" in many devices is necessary.

The p–n junction is one of the key building blocks of all semiconductor electronic devices that ushered in the modern information age. It is rooted in the atomic properties of semiconductors and its development is a clear demonstration of the remarkable power of the fundamental principles of quantum mechanics.

10.9 Problems

10.1 Derive the expression for the Fermi energy for a two-dimensional free electron gas analogous to the corresponding one- and three-dimensional Fermi energies, (10.25) and (10.26), given in the text.

10.2 Use the two-dimensional density-of-states derived in Problem 10.1 above.

(a) Show on the basis of Eq. (10.29) that the chemical potential of a free-electron gas in two dimensions is given by:

$$\mu(T) = k_B T \ln\left[e^{\frac{\pi \hbar^2 N_e}{mk_B T}} - 1\right],$$

for N_e electrons per unit area.

(b) Plot $\mu(T)/E_F$ as a function of kT/E_F as in Figure 10.6(b).

10.3 For a typical 1-D energy band, sketch graphs of the relationships between the wave vector, k, of an electron and its:

(a) energy,

(b) group velocity, and

(c) effective mass.

(d) Sketch the approximate density-of-states $D^{(1)}(E)$ for the energy band of part (a).

10.4 The $E(k_x)$ vs. k_x dependence for an electron in the conduction band of a one-dimensional semiconductor crystal with lattice constant $a = 4\text{Å}$ is given by:

$$E(k_x) = E_2 - (E_2 - E_1)\cos^2[k_x a/2]; \quad E_2 > E_1.$$

(a) Sketch $E(k_x)$ for this band in the reduced and periodic zone schemes.

(b) Find the group velocity of an electron in this band and sketch it as a function of k_x.

(c) Find the effective mass of an electron in this band as a function of k_x and sketch it in the reduced-zone scheme. A uniform electric field E_x is applied in the x direction, in what direction will an electron in the conduction band whose $k_x = 0.2\pi/a$ be accelerated? Repeat for $k_x = 0.5\pi/a$ and $k_x = 0.9\pi/a$. Explain your results.

10.5 Suppose now the corresponding electron energy $E(k_x)$ vs. k_x curve in the valence band is:

$$E(k_x) = -E_3 + E_3 \cos^2[k_x a/2].$$

(a) Sketch $E(k_x)$ for this band in the reduced- and periodic-zone schemes.

(b) Find the group velocity of a hole in this band and sketch it as a function k_x.

(c) Find the effective mass of a hole in this band as a function of k_x and sketch it in the reduced-zone scheme. What is the corresponding effective mass of an electron in the valence band?

(d) A uniform electric field E_x is applied in the x direction. In what direction will a hole in the valence band whose $k_x = 0.2\pi/a$ be accelerated? Repeat for $k_x = 0.5\pi/a$ and $k_x = 0.9\pi/a$. Explain your results.

10.6 Verify the expression for the Fermi level, Eq. (10.48), given in the text.

10.7 A semiconductor has $N_c = 4 \times 10^{17}$ cm^{-3} and $N_v = 6 \times 10^{18}$ cm^{-3} at room temperature and has a band-gap of 1.4 eV. A p–n junction is made in this material with $N_a = 10^{17}$ cm^{-3} on one side, and $N_d = 2 \times 10^{15}$ cm^{-3} on the other side. Assume complete ionization of donors and acceptors.

(a) How many eV separate the Fermi level from the top of the valence band on the p-side at room temperature?

(b) How many eV separate the Fermi level from the bottom of the conduction band on the n-side at room temperature?

(c) What is the built-in voltage across the junction at room temperature?

(d) What is the equilibrium minority carrier (electron) density on the p-side of the junction at room temperature?

(e) By what factor does the minority carrier density on the p-side change when a forward bias of 0.1 eV is applied across the junction? Does it increase or decrease?

11 The density matrix and the quantum mechanic Boltzmann equation

While the dynamic state of a single particle can be specified quantum mechanically in terms of its state function, any rigorous description of the state of a many-particle system would require the complete knowledge of the dynamic state functions of all the particles. That is not always possible. On the other hand, for a large number of particles in, or near, thermal equilibrium in a uniform sample, the principles of statistical mechanics may be invoked to describe the averaged expectation values of the physically observable properties of such a many-particle system. The basic concepts of the density-matrix formalism and the quantum mechanic analog of the classical Boltzmann equation commonly used for optical and magnetic resonance problems of many-particle quantum systems are introduced in this chapter. Applications of this approach to such specific problems as the resonant interaction of electromagnetic radiation with optical media of two-level atoms, nonlinear optics, and the laser rate equations and transient dynamics are discussed in this chapter.

11.1 Definitions of the density operator and the density matrix

Up to this point, in studying the dynamics of quantum mechanic systems, we have assumed that the state of the system can be specified in terms of a precisely known state function $|\Psi\rangle$. On the other hand, for a macroscopic medium containing many particles, it is not always possible to know the exact dynamic states of all the particles in the medium, even for physically identical particles. Often, the most that can be known and specified is a probability distribution function P_Ψ over all the possible states $|\Psi\rangle$ that the N particles per unit volume of the macroscopic medium can be in. The expectation value per unit volume of some physical property represented, for example, by an operator $\hat{Q}$ averaged over the possible states is then:

$$\langle \overline{\hat{Q}} \rangle = N \sum_{|\Psi\rangle} P_\Psi \langle \Psi | \hat{Q} | \Psi \rangle, \tag{11.1}$$

where $|\Psi\rangle$ is assumed to be normalized. The right side of (11.1) in the matrix representation using a set of arbitrarily chosen basis states $|n\rangle$ is:

$$\langle \overline{\hat{Q}} \rangle = N \sum_{m,n} \sum_{|\Psi\rangle} P_\Psi \langle \Psi | m \rangle \langle m | \hat{Q} | n \rangle \langle n | \Psi \rangle. \tag{11.2}$$

For many applications, the representation in which the Hamiltonian of the atoms is diagonal is often used. Since all the factors on the right side of (11.2) are now simple numbers, the order of multiplication is unimportant. The factors can then be rearranged and regrouped:

$$\langle \overline{\hat{Q}} \rangle = N \sum_{m,n} \left[\sum_{|\Psi\rangle} \langle n|\Psi\rangle P_\Psi \langle\Psi|m\rangle \right] \langle m|\hat{Q}|n\rangle$$

and written as:

$$\langle \overline{\hat{Q}} \rangle = N \sum_{m,n} \langle n|\hat{\rho}|m\rangle \, \langle m|\hat{Q}|n\rangle, \tag{11.3}$$

where

$$\langle n|\hat{\rho}|m\rangle \equiv \sum_{|\Psi\rangle} \langle n|\Psi\rangle P_\Psi \langle\Psi|m\rangle \tag{11.4a}$$

can be defined as the element of a matrix called the "density matrix." $\hat{\rho}$ is defined as the corresponding "density operator":

$$\hat{\rho} \equiv \sum_{|\Psi\rangle} |\Psi\rangle P_\Psi \langle\Psi|. \tag{11.4b}$$

Equation (11.3) can in turn be written as the "trace," which in matrix algebra is defined as the sum of the diagonal elements of a matrix, of the product of two operators without referring to any specific representation:

$$\langle \overline{\hat{Q}} \rangle = N \, \text{trace} \, [\, \hat{\rho} \, \hat{Q} \,]. \tag{11.5}$$

11.2 Physical interpretation and properties of the density matrix

The state of a uniform macroscopic medium of many identical particles is now specified in terms of the density matrix, or the density operator $\hat{\rho}$, of the medium. It applies to the situation where the precise dynamic states of the particles in the medium are not necessarily known. What is known is how the particles are distributed statistically over all the states that the particles can possibly occupy. Moreover, it is assumed that the N identical particles in the medium are 'essentially' independent of each other in the sense that the possible states can be specified in terms of the single-particle states $|\Psi\rangle$. It is also assumed that all the particles are subject to two categories of forces. The first is common to all the particles and is represented by the single-particle Hamiltonian $\hat{H}$. The second category corresponds to some weak "randomizing forces" acting differently on different particles due to the possible presence of "relaxation processes."

A simple interpretation of the physical meaning of the elements of the density matrix can be given in terms of, for example, a two-level atom model. In the representation in which the Hamiltonian of the two-level atom $\hat{H}$ is diagonal, the density matrix is a 2x2 matrix:

$$\hat{\rho} = \begin{pmatrix} \langle 1|\hat{\rho}|1\rangle & \langle 1|\hat{\rho}|2\rangle \\ \langle 2|\hat{\rho}|1\rangle & \langle 2|\hat{\rho}|2\rangle \end{pmatrix} \equiv \begin{pmatrix} \rho_{11} & \rho_{12} \\ \rho_{21} & \rho_{22} \end{pmatrix}, \tag{11.6}$$

where the basis states $|1\rangle$ and $|2\rangle$ are the eigen states of the Hamiltonian $\hat{H}$. According to the definition of the density matrix, (11.4a), the diagonal elements of the density matrix are:

$$\rho_{11} = \sum_{\Psi} P_{\Psi} |\langle 1|\Psi\rangle|^2 \quad \text{and} \quad \rho_{22} = \sum_{\Psi} P_{\Psi} |\langle 2|\Psi\rangle|^2. \tag{11.7}$$

They are the probabilities a particle in the state $|\Psi\rangle$ will be found in levels 1 and 2, respectively, averaged over all the states possibly occupied by the particles in the medium.

The off-diagonal elements ρ_{12} and ρ_{21} are the averages of the product of the expansion coefficients $\langle 1|\Psi\rangle$ and $\langle 2|\Psi\rangle$ and they are the complex conjugate of each other:

$$\rho_{12} = \sum_{\Psi} P_{\Psi} \langle 1|\Psi\rangle \langle \Psi|2\rangle = \rho_{21}^*. \tag{11.8a}$$

ρ_{12} and ρ_{21} are in general complex with an averaged product of the amplitudes and relative phase $(\phi_1 - \phi_2)$ of the expansion coefficients:

$$\rho_{12} = \sum_{\Psi} P_{\Psi} \left[|\langle 1|\Psi\rangle| |\langle \Psi|2\rangle| e^{i(\phi_1 - \phi_2)} \right]. \tag{11.8b}$$

If neither one of the diagonal elements is equal to zero, whether the off-diagonal elements vanish or not will depend on whether the relative phase factor $(\phi_1 - \phi_2)$ in (11.8b) averages to zero or not. If the wave functions are "incoherent" with randomly distributed relative phases, the off-diagonal elements as given in (11.8b) are expected to average to zero, even if neither of the expansion coefficients $|\langle 1|\Psi\rangle|$ or $|\langle 2|\Psi\rangle|$ is equal to zero. Thus, given the same population distribution, or diagonal elements, the off-diagonal elements are a measure of the "coherence" of the wave functions of the atoms, or the "atomic coherence" of the atoms in the medium.

The physical significance of the atomic coherence can also be appreciated, especially for optical problems, by considering the macroscopic electric polarization $\vec{P}$ of the two-level atoms in the medium. Take, for example, its z component. According to (11.5), it is:

$$\begin{aligned} P_z = \langle \overline{\hat{P}_z}\rangle &= N \operatorname{trace}\left[\hat{\rho}\,\hat{p}_z\right] \\ &= N \operatorname{trace}\left[\begin{pmatrix} \rho_{11} & \rho_{12} \\ \rho_{21} & \rho_{22} \end{pmatrix} \begin{pmatrix} 0 & p_{z12} \\ p_{z21} & 0 \end{pmatrix} \right] \\ &= N\left(\rho_{12}p_{z21} + \rho_{21}p_{z12} \right), \end{aligned} \tag{11.9}$$

where $\hat{p}_z$ in the present context is the operator representing the z component of the induced electric dipole moment of the individual atoms, not the linear momentum operator! $p_{zmn} \equiv \langle m|\hat{p}_z|n\rangle$ is the corresponding "induced dipole matrix element" of the individual atoms, which we assume to be finite for the problems of interest in the present context. It is assumed that the atoms have no permanent dipole; thus, the diagonal elements of the dipole matrix are zero. (11.9) shows clearly that, if the off-diagonal elements ρ_{12} and ρ_{21} are zero, the macroscopic polarization of the medium also vanishes, even if there are atoms in both level 1 and level 2 and the induced dipole moment of the individual atoms is finite. The off-diagonal elements are then obviously a measure of the phase coherence of the induced oscillating dipoles of the atoms in the medium. The macroscopic polarization of the medium would vanish when the relative phases of the dynamic wave functions of the atoms are randomly distributed.

Extending the above interpretation of the density matrix to the many-level systems in general, *the diagonal elements give the relative populations of the energy levels and the off-diagonal elements give the atomic coherence of the wave functions*. In other words, the populations of the various energy levels N_1, N_2, N_3, ..., N_{mm} are $N\rho_{11}$, $N\rho_{22}$, $N\rho_{33}$, ..., $N\rho_{mm}$. The off-diagonal element ρ_{mn}, where $m \neq n$, gives the coherence of the relative phases of the wave functions between level m and level n.

From the definition of the density matrix and the interpretation of the meaning of its elements given above, it is also possible to derive some of its simple formal properties. For example, from (11.4a):

$$\text{trace } \hat{\rho} = \sum_m \rho_{mm}$$

$$= \sum_{\Psi,m} P_\Psi |\langle \Psi|m\rangle|^2 = \sum_\Psi P_\Psi \langle \Psi|\Psi\rangle$$

$$= \sum_\Psi P_\Psi \equiv 1, \tag{11.10}$$

i.e. the trace of the density matrix or the sum of the diagonal elements of the density matrix must be unity, reflecting the obvious fact that the total population of the N-particle system must be N. Also from (11.4a), the density matrix must be "Hermitian," meaning:

$$\rho_{mn} = \sum_\Psi P_\Psi \langle m|\Psi\rangle \langle \Psi|n\rangle = \sum_\Psi P_\Psi^* \langle n|\Psi\rangle^* \langle \Psi|m\rangle^* = \rho_{nm}^*. \tag{11.11}$$

Finally, if the N-particle system is in a "pure state," meaning all the particles in the medium are in the same particular state, Ψ_0, or:

$$P_\Psi = \delta_{\Psi\Psi_0}, \tag{11.12}$$

then

$$
\hat{\rho}_0^2 = \left(\sum_{\Psi} P_{\Psi} |\Psi\rangle \langle\Psi| \right) \left(\sum_{\Psi'} P_{\Psi'} |\Psi'\rangle \langle\Psi'| \right)
$$

$$
= \left(\sum_{\Psi} \delta_{\Psi\Psi_0} |\Psi\rangle \langle\Psi| \right) \left(\sum_{\Psi'} \delta_{\Psi'\Psi_0} |\Psi'\rangle \langle\Psi'| \right)
$$

$$
= |\Psi_0\rangle \langle\Psi_0|\Psi_0\rangle \langle\Psi_0| = |\Psi_0\rangle \langle\Psi_0|
$$

$$
= \hat{\rho}_0. \tag{11.13}
$$

11.3 The density matrix equation or the quantum mechanic Boltzmann equation

Since the state of the N-particle system is now specified by the density operator $\hat{\rho}$ or the density matrix ρ_{mn}, its dynamics are completely characterized by the time-dependence of $\hat{\rho}$ or ρ_{mn}, as determined by the equation-of-motion of $\hat{\rho}$ or ρ_{mn}.

From the definition of the density operator, (11.4b), $\hat{\rho}$ can vary with time through two sources. One is the explicit time-dependence of the probability distribution function $P_{\Psi}(t)$ independent of the time dependence of the state $|\Psi\rangle$. The second is the implicit time-dependence because the state function $|\Psi(t)\rangle$ is time-dependent. Thus, the time rate of change of $\hat{\rho}$ is given by:

$$
\frac{\mathrm{d}}{\mathrm{d}t}\hat{\rho} = \sum_{\Psi} \left[\left(\frac{\partial}{\partial t} |\Psi(t)\rangle \right) P_{\Psi}\langle\Psi| + |\Psi\rangle \left(\frac{\partial}{\partial t} P_{\Psi}(t) \right) \langle\Psi| \right.
$$

$$
\left. + |\Psi\rangle P_{\Psi} \left(\frac{\partial}{\partial t} \langle\Psi(t)| \right) \right]. \tag{11.14}
$$

Let us consider first the case when there is no relaxation process involved, or there are no randomizing forces acting differently on different particles, and all the particles in the sample have exactly the same Hamiltonian $\hat{H}$; therefore, according to Schrödinger's equation:

$$
\frac{\partial}{\partial t} |\Psi(t)\rangle = -\frac{\mathrm{i}}{\hbar} \hat{H}|\Psi(t)\rangle \text{ and } \frac{\partial}{\partial t} \langle\Psi(t)| = \frac{\mathrm{i}}{\hbar} \langle\Psi(t)|\hat{H}. \tag{11.15a}
$$

Moreover, if there are no random forces acting on the particles and all the particles see identical forces, whatever particles that are in the same state $|\Psi(t)\rangle$ at the time t must be in the same state $|\Psi(t + \Delta t)\rangle$ at $t + \Delta t$. Similarly, any particles starting out in different states at t cannot end up in the same state at $t + \Delta t$. This is because the state functions for all the particles satisfy the same time-dependent Schrödinger equation and its solution is unique. That is, no two states satisfying the same initial condition at t can lead to different solutions at $t + \Delta t$. Similarly, no two states with different initial conditions at t can end up with the same solution at $t + \Delta t$. In other words, as the state function $|\Psi(t)\rangle$ evolves in time in $|\Psi\rangle$-space, no two trajectories can cross, if the Hamiltonian for all the particles are the same. Therefore, in the absence of relaxation processes, the probability distribution function P_Ψ can not change explicitly with time, or:

$$\frac{\partial}{\partial t} P_\Psi = 0. \tag{11.15b}$$

Substituting (11.15a) and (11.15b) into (11.14) leads to the equation of motion for $\hat{\rho}$ in the absence of relaxation processes:

$$\frac{d}{dt} \hat{\rho} = -\frac{i}{\hbar} [\hat{H}, \hat{\rho}]. \tag{11.16}$$

Note that, although Eq. (11.16) looks somewhat like Heisenberg's equation of motion (2.49), there is a sign difference in the front. Very significantly, the equation of motion for $\hat{\rho}$ in the form of (11.16) is based on Schrödinger's equation of motion; therefore, it is based on Schrödinger's picture, not Heisenberg's picture, of the dynamics of the many-particle system.

We will now take into account the relaxation processes, which are always present in real macroscopic media. Because the randomizing forces corresponding to the relaxation processes act differently on different particles, the condition (11.15b) no longer holds; therefore, there must be an additional term in the rate of change of the density operator:

$$\frac{d}{dt} \hat{\rho} = -\frac{i}{\hbar} [\hat{H}, \hat{\rho}] + \left(\frac{\partial \hat{\rho}}{\partial t}\right)_{\text{Random}}. \tag{11.17}$$

This term, $(\partial \hat{\rho}/\partial t)_{\text{Random}}$, cannot be specified on the basis of first principles, because there is no way of knowing exactly the random forces corresponding to the relaxation processes acting on all the individual particles. It can only be approximated phenomenologically. The usual argument is that, if the medium is not too far from thermal equilibrium, the corresponding rate of change will be proportional to the deviation of the density matrix element $\rho_{mn} \equiv \langle m|\hat{\rho}|n\rangle$ from its thermal equilibrium value $\rho_{mn}^{(\text{th})}$:

$$\left(\frac{\partial \rho_{mn}}{\partial t}\right)_{\text{Random}} = -\frac{\rho_{mn} - \rho_{mn}^{(\text{th})}}{T_{mn}}, \tag{11.18}$$

where the proportionality constant T_{mn}^{-1} is a phemenological relaxation rate and $T_{mn}^{-1} = T_{nm}^{-1}$, in the basis in which $\hat{H}$ is diagonal or $\hat{H}|m\rangle = E_m|m\rangle$. Combining Eqs. (11.17) and (11.18) leads to the very important phenomenological equation of motion for the density matrix:

$$\frac{d}{dt} \hat{\rho}_{mn} = -\frac{i}{\hbar} [\hat{H}, \hat{\rho}]_{mn} - \frac{\rho_{mn} - \rho_{mn}^{(\text{th})}}{T_{mn}}. \tag{11.19}$$

Equation (11.19) for the many-particle system is analogous to the time-dependent Schrödinger equation for the single particle with the addition of the relaxation process. It is also the quantum mechanical analog of the Boltzmann equation for the distribution function in the six-dimensional phase-space of the position coordinates and the velocity of the particles according to classical statistical mechanics. Such a quantum mechanic Boltzmann equation is widely used to study, for example, the interaction of coherent electromagnetic waves with macroscopic optical media.

11.4   Examples of the solutions and applications of the density-matrix equations

To see how the density-matrix approach is used in optical problems, let us consider some examples of the solutions of the density-matrix equations, ranging from the simple case of no relaxation process and no external perturbation acting on the atoms in the medium to the far more complicated cases of nonlinear response of optical materials to intense laser light and the transient dynamics of different types of lasers.

Solution of the density-matrix equation in the absence of relaxation processes and external perturbation

In the absence of any relaxation process, the density matrix equation is simply, from (11.16):

$$\frac{\mathrm{d}}{\mathrm{d}t} \rho_{mn} = -\frac{\mathrm{i}}{\hbar} [\hat{H}, \hat{\rho}]_{mn}.$$

If there is also no perturbation, it is assumed that the time-independent Schrödinger equation:

$$\hat{H}|m\rangle = E_m|m\rangle \qquad (11.20)$$

for the dynamic system can be solved; i.e. the eigen values and eigen functions of the Hamiltonian $\hat{H}$ in the above density matrix equation are assumed known. Thus,

$$\frac{\mathrm{d}}{\mathrm{d}t} \rho_{mn} = -\mathrm{i}\omega_{mn} \rho_{mn}, \qquad (11.21)$$

where $\omega_{mn} \equiv (E_m - E_n)/\hbar$. The solution of (11.21) is trivial:

$$\rho_{mn}(t) = \rho_{mn}(0)\, \mathrm{e}^{-\mathrm{i}\omega_{mn}t}; \qquad (11.22)$$

it gives, however, a physically unrealistic result. Whatever the initial populations of the various energy levels are, they will never change and return to thermal equilibrium. Also, any initial atomic coherence and, hence, macroscopic polarization will persist forever and never vanish.

We know, of course, that any medium initially in a non-equilibrium state at a temperature T must eventually return to thermal equilibrium, in which the population distribution is the Boltzmann distribution and the atomic coherence vanishes. In other words, the state of the medium must eventually approach the thermal equilibrium state characterized by the thermal equilibrium density matrix:

$$\rho_{mn}^{(\text{th})} = \delta_{mn} \frac{e^{-E_m/k_B T}}{\sum\limits_m e^{-E_m/k_B T}}. \tag{11.23}$$

Solution of the density-matrix equation with relaxation processes

To ensure that the solutions of the density-matrix equation reflect this physical reality, the equation of motion must include a relaxation term of the form (11.18). Solving the resulting equation, (11.19):

$$\frac{d}{dt}\rho_{mn} = -i\omega_{mn}\rho_{mn} - \frac{\rho_{mn} - \rho_{mn}^{(\text{th})}}{T_{mn}}, \tag{11.24}$$

one obtains:

$$\rho_{mn}(t) = \begin{cases} \rho_{mm}^{(\text{th})} + \left[\rho_{mm}(0) - \rho_{mm}^{(\text{th})}\right] e^{-t/T_{mm}} & \text{for} \quad m = n \\ \rho_{mn}(0)\, e^{-(i\omega_{mn} + \frac{1}{T_{mn}})\,t} & \text{for} \quad m \neq n. \end{cases} \tag{11.25}$$

Equation (11.25) shows that the diagonal terms of the density matrix will always relax to the Boltzmann distribution with the relaxation time T_{mm}, which is referred to in the literature on photonics as the population relaxation time. In the literature on magnetic-resonance phenomena, it is often referred to as the "longitudinal relaxation time" or the "T_1-time." Similarly, the off-diagonal terms will always eventually vanish at the relaxation rate T_{mn}^{-1}. In optics literature, T_{mn} is often referred to as the "atomic coherence time"; in the literature on magnetic-resonance phenomena, it is also referred to as the "transverse relaxation time" or the "T_2-time." The terminology T_1 and T_2 used in magnetic resonance work is often adopted and also used in optical problems. Note that for the same pair of m and n values, the corresponding T_2 time is always shorter than twice the T_1 time because of energetic considerations: it is always easier to change the phase than change the populations in a relaxation process. The fact that the off-diagonal elements and diagonal elements of the density matrix can relax with different rates is of great practical significance in optics.

Density-matrix equation with perturbation

Suppose now the medium is also subject to a time-dependent external perturbation represented by the operator $\hat{V}(t)$ in the total Hamiltonian:

$$\hat{H} = \hat{H}_0 + \hat{V}(t), \tag{11.26}$$

where $\hat{H}_0$ is the unperturbed Hamiltonian in the absence of any external perturbation. Both $\hat{H}_0$ and $\hat{V}(t)$ are common to all the atoms in the medium. In the representation in which $\hat{H}_0$ is diagonal, the corresponding density-matrix equation is, from (11.19):

$$\frac{d}{dt}\rho_{mn} + i\omega_{mn}\rho_{mn} + \frac{\rho_{mn} - \rho_{mn}^{(th)}}{T_{mn}} = \frac{i}{\hbar}\sum_{m'}\left[\rho_{mm'}V(t)_{m'n} - V(t)_{mm'}\rho_{m'n}\right]. \qquad (11.27)$$

It is assumed that the basis states are still the eigen states of $\hat{H}_0$, and the form of the phenomenological relaxation term (11.18) remains the same even in the presence of the perturbation. This is a good approximation, *if the time-dependent perturbation is weak and varies rapidly compared to the relaxation rates*. Equation (11.27) can be used to study a great variety of optical problems ranging from, for example, the linear absorption and dispersion effects in optical media to the transient dynamics of lasers and many of the nonlinear optical effects. We will now look at a few of such examples.

Interaction of electromagnetic radiation with an optical medium of two-level atoms

A simple example of the application of the density-matrix equation of the form of (11.27) to optics problems is the resonant interaction of an electromagnetic wave with an optical medium consisting of identical two-level atoms. For electric dipole inter-action, the perturbation term $\hat{V}$ in the Hamiltonian:

$$\hat{H} = \hat{H}_0 + \hat{V}(t) \qquad (11.28)$$

is of the form:

$$\hat{V}(t) = -\hat{\vec{p}} \cdot \vec{E}(t) = e\hat{\vec{r}} \cdot \vec{E}(t), \qquad (11.28a)$$

where $\hat{\vec{p}}$ is the operator representing the electric dipole of the atom. For a monochromatic transverse linearly polarized wave, the electric field acting on the medium is assumed to be of the form:

$$\vec{E}(t) = \left[\tilde{E}_z e^{-i\omega_0 t} + \tilde{E}_z^* e^{i\omega_0 t}\right]\mathbf{e}_z, \qquad (11.29)$$

where $\mathbf{e}_z$ is the unit vector in the z direction of the Cartesian coordinates. With a suitably chosen time origin $t = 0$, $E_z = E_z^*$ can be taken to be purely real. We assume that the wavelength of the electromagnetic wave is much larger than the size of the macroscopic sample under study and the inter-atomic distances in the medium. The amplitude of the electric field E_z can, therefore, be assumed to be a spatially independent known constant parameter over the volume of the medium being considered. Thus, the matrix element of the perturbation term in (11.27) in the representation in which $\hat{H}_0$ is diagonal is of the form:

$$V(t)_{mn} = e\langle m|z|n\rangle\left[\tilde{E}_z e^{-i\omega_0 t} + \text{complex conjugate}\right] \equiv ez_{mn}\left[\tilde{E}_z e^{-i\omega_0 t} + c.c.\right]. \quad (11.30)$$

For two-level atoms or ions with no permanent dipole moment, $V(t)_{mn}$ is simply a 2x2 purely off-diagonal matrix of the form:

$$\hat{V}(t) = \begin{pmatrix} 0 & V(t)_{12} \\ V(t)_{21} & 0 \end{pmatrix}. \tag{11.31}$$

The relevant density matrix characterizing the state of the macroscopic medium is also a simple 2x2 matrix:

$$\hat{\rho} = \begin{pmatrix} \rho_{11} & \rho_{12} \\ \rho_{21} & \rho_{22} \end{pmatrix}. \tag{11.32}$$

Since the density matrix is Hermitian and its trace is equal to 1 according to (11.10) and (11.11), of the four matrix elements, there are only two independent variables, for example, $(\rho_{11} - \rho_{22}) = 2\rho_{11} - 1$ and $\rho_{12} = \rho_{21}^*$. For optical problems, the physical variables of interest are the population difference of the two levels, $N_1 - N_2$, and the macroscopic polarization of the medium P_z:

$$N_1 - N_2 \equiv N(\rho_{11} - \rho_{22}), \tag{11.33a}$$

$$P_z = \overline{\langle \hat{P}_z \rangle} = N[\rho_{12}(-ez_{21}) + \rho_{21}(-ez_{12})]. \tag{11.33b}$$

Of the four density-matrix equations given by (11.27) for the two-level system being considered, there are, therefore, only two independent coupled differential equations of interest:

$$\frac{\mathrm{d}}{\mathrm{d}t}(\rho_{11} - \rho_{22}) + \frac{1}{T_1}\left[(\rho_{11} - \rho_{22}) - (\rho_{11}^{(\mathrm{th})} - \rho_{22}^{(\mathrm{th})})\right] = \frac{2\mathrm{i}}{\hbar}\left[\rho_{12}V_{21} - V_{12}\rho_{21}\right]. \tag{11.34a}$$

$$\left[\frac{\mathrm{d}}{\mathrm{d}t} + \mathrm{i}\omega_{21} + \frac{1}{T_2}\right]\rho_{21} = -\frac{\mathrm{i}}{\hbar}(\rho_{11} - \rho_{22})V_{21}, \tag{11.34b}$$

where the convention of naming the relaxation times for the diagonal and off-diagonal elements as T_1 and T_2 is used. In optical problems, the relaxation rates are always much less than the transition frequency: $\frac{1}{2T_1} \leq \frac{1}{T_2} \ll \omega_{21}$; the equality applies if there is no dephasing relaxation process other than that associated with the population relaxation process. The factor of 2 in the inequality above is due to the fact that T_1^{-1} refers to relevant wave function *squared* or the *population* decay, while T_2^{-1} refers to the off-diagonal element of the density matrix which involves the complex *amplitude* of the relevant wave functions.

   In principle, given the initial conditions on the variables $(\rho_{11} - \rho_{22})$ and $\rho_{21} = \rho_{12}^*$, the transient and steady state responses of the medium to an applied electromagnetic wave can be found by solving the coupled equations (11.34a) and (11.34b). In practice, it is in general an impossible task to carry out without extensive approximations, mainly because the time-dependence in the perturbation term $\hat{V}(t)$ makes these coupled differential equations with time-varying coefficients. One common approach is to look for response of the medium at the optical frequency ω_0 near the transition frequency, $\omega_0 \approx \omega_{21}$, by making use of the basic technique of time-dependent perturbation theory. A more systematic development of the theory will be given in the following subsection on nonlinear optics. Here, we will follow the usual approach in linear optics theory first.

Equation (11.34b) shows that lowest order effect of turning on a weak applied electric field E_z is to induce an atomic coherence ρ_{12} that is linearly proportional to the field. Equation (11.34a) shows that the corresponding change in the population difference will be proportional to the square of the applied field E_z. Thus, *for linear optics*, we can approximate first the population difference, $(\rho_{11} - \rho_{22})$, in (11.34b) by the unperturbed population distribution, or $(\rho_{11} - \rho_{22}) \cong (\rho_{11}^{(0)} - \rho_{22}^{(0)})$ in the steady state, and from (11.30) and (11.34b):

$$\left[\frac{d}{dt} + i\omega_{21} + \frac{1}{T_2}\right] \rho_{21} = -\frac{iez_{21}}{\hbar} \left(\rho_{11}^{(0)} - \rho_{22}^{(0)}\right) \left[\tilde{E}_z e^{-i\omega_0 t} + c.c.\right]. \tag{11.35}$$

The steady-state solution of (11.35) will contain terms with time variations $e^{-i\omega_0 t}$ and $e^{+i\omega_0 t}$:

$$\rho_{21}(t) \cong -\frac{\left(\rho_{11}^{(0)} - \rho_{22}^{(0)}\right) ez_{21} \tilde{E}_z}{\hbar} \left[\frac{e^{-i\omega_0 t}}{(\omega_{21} - \omega_0) - i/T_2} + \frac{e^{+i\omega_0 t}}{(\omega_{21} + \omega_0) - i/T_2}\right]. \tag{11.36}$$

Near the resonance, $\omega_0 \approx \omega_{21} \gg 1/T_2$, the magnitude of the complex amplitude of the resonant term varying as $e^{-i\omega_0 t}$ is much larger than that of the anti-resonant term varying as $e^{+i\omega_0 t}$; thus, the atomic coherence near the resonance can be approximated by:

$$\rho_{21}(t) \equiv \tilde{\rho}_{21} e^{-i\omega_0 t} \approx -\frac{\left(\rho_{11}^{(0)} - \rho_{22}^{(0)}\right) ez_{21} \tilde{E}_z}{\hbar} \left[\frac{e^{-i\omega_0 t}}{(\omega_{21} - \omega_0) - i/T_2}\right], \tag{11.37a}$$

where $\tilde{\rho}_{21}$ is by definition its complex amplitude. From (11.11), we have:

$$\rho_{12}(t) \equiv \tilde{\rho}_{12} e^{+i\omega_0 t} \approx -\frac{\left(\rho_{11}^{(0)} - \rho_{22}^{(0)}\right) ez_{12} \tilde{E}_z^*}{\hbar} \left[c \frac{e^{+i\omega_0 t}}{(\omega_{21} - \omega_0) + i/T_2}\right]. \tag{11.37b}$$

Let us now examine some of the consequences of these results. Equations (11.37a) and (11.37b) lead directly to one of the most important parameters in linear optics – the "complex susceptibility" $\chi(\omega_0)$ of an optical medium, which is by definition the ratio of the complex amplitude of the induced macroscopic polarization to the Maxwell field in the medium E_z:

$$P_z(t) = \chi_{zz}(\omega_0) \tilde{E}_z e^{-i\omega_0 t} + \chi_{zz}^*(\omega_0) \tilde{E}_z^* e^{i\omega_0 t}. \tag{11.38}$$

From (11.38), (11.37a), (11.37b), (11.33a), and (11.33b), we have:

$$\chi_{zz}(\omega_0) \cong \frac{\left(N_1^{(0)} - N_2^{(0)}\right) e^2 |z_{12}|^2}{\hbar} \left[\frac{(\omega_{21} - \omega_0)}{(\omega_{21} - \omega_0)^2 + T_2^{-2}} + \frac{\mathrm{i}\, T_2^{-1}}{(\omega_{21} - \omega_0)^2 + T_2^{-2}}\right].$$

$$(11.39)$$

It shows explicitly that the line widths of the absorption and dispersion curves are determined by the atomic coherence time T_2, not the population relaxation time T_1. Note that the real and imaginary parts of the *linear* complex susceptibility, (11.39), satisfy the well-known Kramers–Kronig relations:

$$\chi'(\omega) = \frac{1}{\pi}\, P \int_{-\infty}^{\infty} \frac{\chi''(\omega')}{\omega' - \omega}\, \mathrm{d}\omega',$$

$$\chi''(\omega) = \frac{1}{\pi}\, P \int_{-\infty}^{\infty} \frac{\chi'(\omega')}{\omega' - \omega}\, \mathrm{d}\omega', \qquad (11.40)$$

where P stands for the Cauchy principal value of the integral that follows. (See, for example, Yariv (1989), Appendix 1.) The corresponding linear complex dielectric constant $\varepsilon(\omega_0)$ of the medium is:

$$\varepsilon_{zz}(\omega_0) = \varepsilon_{zz}'(\omega_0) + \mathrm{i}\varepsilon_{zz}''(\omega_0) = \varepsilon_0 + 4\pi\chi_{zz}(\omega_0), \qquad (11.41)$$

where ε_0 is the dielectric constant of the host medium in which the two-level atoms are imbedded. Thus, the real and imaginary parts of the complex dielectric constant are, respectively, of the forms:

$$\varepsilon_{zz}'(\omega_0) = \varepsilon_0 + \Delta\varepsilon \left[\frac{x_{21} - x_0}{1 + (x_{21} - x_0)^2}\right] \qquad (11.42a)$$

and

$$\varepsilon_{zz}''(\omega_0) = \Delta\varepsilon \left[\frac{1}{1 + (x_{21} - x_0)^2}\right], \qquad (11.42b)$$

where

$$\Delta\varepsilon \equiv \frac{4\pi \left(N_1^{(0)} - N_2^{(0)}\right) e^2 |z_{12}|^2 T_2}{\hbar}, \quad x_{21} \equiv \omega_{21}\, T_2, \quad \text{and} \quad x_0 \equiv \omega_0\, T_2. \qquad (11.43)$$

These are some of the most fundamental results in optics, which are now derived here rigorously quantum mechanically. The real part of the complex dielectric constant $\varepsilon'(\omega_0)$, (11.42a), gives the well-known dispersion characteristic of such a medium and is shown schematically in Figure 11.1(a). The imaginary part,(11.42b), gives the equally well-known absorption characteristic of the medium and has the characteristic Lorentzian line shape; it is shown schematically in Figure 11.1(b).

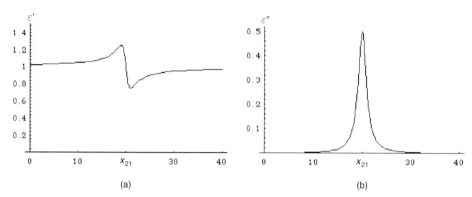

Figure 11.1 Examples of (a) the dispersion and (b) the absorption characteristics of an optical medium of two-level atoms $(\varepsilon_0 = 1, \Delta\varepsilon = 0.5,$ and $x_{21} = 20$; see (11.42a) and (11.42b)).

The intensity attenuation coefficient for a monochromatic electromagnetic wave propagating in the medium can be obtained from (11.42b). It is:

$$\alpha_{zz} \equiv \frac{\varepsilon_{zz}''(\nu_0)\omega_0}{\sqrt{\varepsilon_0}c} \cong \frac{4\pi^2\,(N_1 - N_2)\,\nu_0\,e^2|z_{12}|^2}{\sqrt{\varepsilon_0}\hbar c}g_f(\nu_0), \tag{11.44}$$

where

$$g_f(\nu_0) \equiv \left[\frac{1}{\pi}\frac{(2\pi T_2)^{-1}}{(\nu_0 - \nu_{21})^2 + (2\pi T_2)^{-2}}\right] \tag{11.45}$$

is the "fluorescence line shape function" as in (8.30). Equation (11.44) is completely equivalent to the results obtained in Section 8.6, (8.30)–(8.32), thus justifying the ad hoc simple single-atom model used there to derive those results.

For a medium in thermal equilibrium, when the intensity is so weak that the population change due to the induced transition as represented by the right side of (11.34a) is negligible compared with the relaxation rate T_1^{-1}, the population difference is essentially the thermal equilibrium distribution: $(\rho_{11} - \rho_{22}) \cong (\rho_{11}^{(0)} - \rho_{22}^{(0)}) \cong (\rho_{11}^{(th)} - \rho_{22}^{(th)})$. When the radiation is more intense and the rate of induced transition is not negligible compared to the population relaxation rate T_1^{-1}, substituting (11.37a) into (11.34a) gives the steady-state population difference in the presence of the wave as a function of the incident intensity $I = \frac{|\bar{E}|_z^2 c}{2\pi}$:

$$(N_1 - N_2) = \frac{\left(N_1^{(th)} - N_2^{(th)}\right)}{1 + I/I_{\text{sat}}}, \tag{11.46}$$

where

$$I_{\text{sat}}^{-1} \equiv \frac{8\pi\,T_1\,e^2|z_{12}|^2}{\hbar^2 c}\left[\frac{T_2^{-1}}{(\omega_{21} - \omega_0)^2 + T_2^{-2}}\right]. \tag{11.47}$$

I_{sat} is the so-called "saturation parameter" for the induced resonance transition between the two atomic levels. The leading term in an expansion of the result given in (11.46) in powers of I/I_{sat} gives:

$$(N_1 - N_2) - (N_1^{(th)} - N_2^{(th)}) \cong - \frac{4T_1 \, e^2|z_{12}|^2|\tilde{E}_z|^2(N_1^{(th)} - N_2^{(th)})}{\hbar^2}$$

$$\left[\frac{T_2^{-1}}{(\omega_{21} - \omega_0)^2 + T_2^{-2}} \right]. \tag{11.48}$$

Equation (11.48) gives the well-known result for induced resonance transition between two atomic levels consistent with (8.30)–(8.32). Equation (11.46) shows that there is an appreciable change in the population difference only when the intensity of the incident radiation is not negligible compared with the saturation parameter. When the intensity is much greater than the saturation parameter, the populations of the two levels become equalized.

Optical Bloch equations

An alternative formulism in dealing with the optical properties of macroscopic media of two-level atoms is to use the variables the population difference, $N_1 - N_2$, and the macroscopic polarization, P_z, as defined in (11.33a) and (11.33b), respectively, instead of the diagonal and off-diagonal elements of the density-matrix as in the previous subsection. The rate equation of the population difference $N_1 - N_2$ follows immediately from (11.33a), (11.33b), and (11.34a), and it is, for $\omega \approx \omega_{21}$:

$$\frac{d}{dt}(N_1 - N_2) + \frac{1}{T_1}\left[(N_1 - N_2) - \left(N_1^{(th)} - N_2^{(th)}\right)\right] = -\frac{4\tilde{E}_z}{\hbar}\,\mathrm{Im}(\tilde{P}_z), \tag{11.49}$$

where $\tilde{p}_z$ is the complex amplitude of P_z, defined as follows:

$$P_z \equiv \tilde{P}_z e^{-i\omega t} + \tilde{P}_z^* e^{+i\omega t}.$$

One can also obtain from (11.33a), (11.33b), (11.34a), and (11.34b) second-order differential equation for P_z, after some algebra:

$$\frac{d^2}{dt^2} P_z + \frac{2}{T_2}\frac{d}{dt} P_z + \omega_{21}^2 P_z = \frac{2\omega_{21}\, e^2\, |z_{21}|^2}{\hbar}(N_1 - N_2)\left[\tilde{E}_z e^{-i\omega t} + c.c.\right] \tag{11.50}$$

for $\omega \approx \omega_{21} \gg 1/T_2$. To the first-order of approximation of $ez_{21}\tilde{E}_z$ in (11.49) and (11.50), $\mathrm{Im}\tilde{P}_z \propto \tilde{E}_z$ and $(N_1 - N_2) \cong (N_1^{(th)} - N_2^{(th)})$ in the steadystate; (11.50) is then approximately:

$$\frac{d^2}{dt^2} P_z + \frac{2}{T_2}\frac{d}{dt} P_z + \omega_{21}^2 P_z \cong \frac{2\omega_{21}\, e^2\, |z_{21}|^2}{\hbar}\left(N_1^{(th)} - N_2^{(th)}\right)\left[\tilde{E}_z e^{-i\omega t} + c.c.\right], \tag{11.50a}$$

which has the form of a driven damped classical harmonic oscillator equation. It is in a form that is particularly useful for comparing the results in linear optics obtained on the basis of the classic harmonic oscillator model with those obtained from the quantum mechanical model, such as (11.49) or (11.50a).

Equations. (11.49) and (11.50) are known as the optical Bloch equations. They are often used to characterize the optical properties of macroscopic media of two-level atoms instead of the using the density-matrix equations. In the density-matrix formalism, one solves the dynamic equations for the density-matrix elements first and then converts the results to the physically more meaningful parameters $N_1 - N_2$ and P_z. In the optical Bloch formalism, one finds the appropriate Bloch equations for the macroscopic physical parameters of interest first and then solves the equations for these parameters. The two approaches are completely equivalent. The former approach tends to bring out the analogies between the optical and magnetic resonance phenomena more readily.

Nonlinear optical susceptibilities

A most important class of new optical phenomena that can be studied with intense laser light is in the realm of nonlinear optics. (See, for example, Shen 1984).) The density-matrix equation (11.27) shows that the response of the medium to an applied electromagnetic wave can lead to multiple frequency components beyond those contained in the original incident wave. This is because the time-dependent perturbation term $V(t)_{mn}$ appears in the coefficients of the coupled differential equations for the diagonal and off-diagonal elements. Thus, any initial frequency components in $V(t)_{mn}$ will generate multiples and mixtures of the original frequency components in the solutions of the density matrix equations. In practice, these terms can lead to a wide range of applications based on the possibility of generating harmonics and sum- and difference-frequency components of the spectral components contained in the incident wave and in the resonances in the material. These effects are characterized by the "nonlinear optical susceptibilities" of the medium. We will now develop a simple theory of such effects based on the steady-state solutions of Eq. (11.27), without any weak-field assumption restricting the response of the medium as in (11.35).

To find a more general steady solution of the density-matrix equation (11.27), it can be put in the form of an integral equation:

$$\rho_{mn}(t) = \int_{-\infty}^{t} \left\{ \frac{\rho_{mn}^{(th)}}{T_{mn}} + \frac{i}{\hbar} \sum_{m'} \left[\rho_{mm'}(t')V(t')_{m'n} - V(t')_{mm'}\rho_{m'n}(t') \right] \right\}$$
$$e^{(i\omega_{mn}+1/T_{mn})(t'-t)}dt'. \tag{11.51}$$

It is in a form that is more convenient to apply the time-dependent perturbation technique to. The two terms in the bracket $\{\ldots\}$ on the right side of (11.51) correspond to the relaxation term in the absence of the field and the induced response of the medium due to the presence of the electromagnetic wave, respectively.

The exponential factor multiplying these terms corresponds to the Green's function of the differential equation corresponding to the left side of (11.27). If the perturbation term corresponding to the field-induced response is negligible compared to the relaxation term, (11.51) shows that the unperturbed density matrix, which we designate as $\rho_{mn}^{(0)}$, would be equal to the thermal equilibrium distribution:

$$\rho_{mn}^{(0)} = \rho_{mn}^{(\text{th})} = \delta_{mn} \frac{e^{-E_m/k_{\text{B}}T}}{\sum_m e^{-E_m/k_{\text{B}}T}}. \tag{11.52}$$

In the presence of a small perturbation due to the applied electric field, we assume that the density matrix can be expanded in a power series in successive orders of V_{mn}:

$$\rho_{mn} = \rho_{mn}^{(0)} + \rho_{mn}^{(1)} + \rho_{mn}^{(2)} + \ldots + \rho_{mn}^{(n)}. \tag{11.53}$$

Substituting (11.53) into (11.51) and equating terms of the same order give the general nth-order perturbation solution of (11.27) or (11.51) in the integral form:

$$\rho_{mn}^{(n)}(t) = \int_{-\infty}^{t} \left\{ \frac{i}{\hbar} \sum_{m'} \left[\rho_{mm'}^{(n-1)}(t') V(t')_{m'n} - V(t')_{mm'} \rho_{m'n}^{(n-1)}(t')(t') \right] \right\}$$
$$e^{(i\omega_{mn}+1/T_{mn})(t'-t)} dt', \tag{11.54}$$

for $n = 1, 2, 3, \ldots$. Thus, with the known zeroth-order term, (11.52), once the perturbation term involving the applied E-field is specified, repeated straightforward integration of (11.54) to successively higher orders will give the density matrix to any arbitrary order of the E-field.

As an example of the application of this routine, consider again the problem considered above of a monochromatic wave incident on a medium of two-level atoms. First, (11.52) shows that the unperturbed density matrix is purely diagonal. Therefore, the atomic coherence in thermal equilibrium is zero and the diagonal elements give the Boltzmann distribution, or whatever the initial unperturbed population-difference is in the steady state. For electric-dipole interaction, the perturbation term is a purely off-diagonal matrix of the form (11.31). Substituting the corresponding perturbation term (11.28a)–(11.31) and $\rho_{mn}^{(0)}$ into (11.54) immediately gives, for example:

$$\rho_{mn}^{(1)} = -\frac{\left(\rho_{nn}^{(0)} - \rho_{mm}^{(0)}\right)}{\hbar} \left[\frac{e\, z_{mn} \tilde{E}_z e^{-i\omega_0 t}}{(\omega_{mn} - \omega_0) - iT_{mn}^{-1}} + \frac{e\, z_{mn} \tilde{E}_z^* e^{i\omega_0 t}}{(\omega_{mn} + \omega_0) - iT_{mn}^{-1}} \right]. \tag{11.55}$$

Specializing (11.55) to the case $m = 2$ and $n = 1$, it gives the same result as (11.36):

$$\rho_{21}^{(1)}(t) \cong -\frac{\left(\rho_{11}^{(0)} - \rho_{22}^{(0)}\right) e z_{21} \tilde{E}_z}{\hbar} \left[\frac{e^{-i\omega_0 t}}{(\omega_{21} - \omega_0) - i/T_2} + \frac{e^{+i\omega_0 t}}{(\omega_{21} + \omega_0) - i/T_2} \right].$$

Continuing on to the second-order term in (11.54), the corresponding density matrix will be purely diagonal and gives exactly the population change as in (11.48).

In the general case involving multi-level atoms, the second-order terms will also contain off-diagonal terms that vary at the second-harmonic frequency $2\omega_0$ as well as a time-independent term. These atomic coherence terms will give rise to induced macroscopic polarization terms $\vec{P}^{(2)}(\omega_0 + \omega_0 = 2\omega_0)$ and $\vec{P}^{(2)}(\omega_0 - \omega_0 = 0)$ and to the corresponding nonlinear optical susceptibilities $\tilde{\bar{\chi}}^{(2)}(\omega_0 + \omega_0 = 2\omega_0)$ and $\tilde{\bar{\chi}}^{(2)}(\omega_0 - \omega_0 = 0)$, respectively, which are third-rank tensors. An oscillating macroscopic polarization at $2\omega_0$ will radiate coherent electromagnetic radiation at this frequency and give rise to the phenomenon of second harmonic generation in the nonlinear optical medium. Similarly, the d.c. term in the induced macroscopic polarization will lead to a d.c. electric field, which is known as the optical-rectification effect in the nonlinear medium. It is clear that the higher order terms in the general solution, (11.54), of the density-matrix equation represent a great variety of nonlinear optical effects, some of which have already been observed experimentally and studied extensively with the help of a variety of lasers. Many more remain to be discovered and studied. The field of nonlinear optics is a rich and active field of research in modern optics. It is another example of the triumphs of the basic principles of quantum mechanics.

Laser rate equations and transient oscillations

The dynamic response of optical media to fields with slowly varying complex amplitudes of the form, for example:

$$\vec{E}(t) \equiv \left[\tilde{E}_z(t)\,e^{-i\omega t} + \tilde{E}_z^*(t)\,e^{i\omega t}\right]\mathbf{e}_z$$

can also be analyzed on the basis of the density-matrix equations. The transient and stability characteristics of lasers are such cases. Using again the two-level atom model for the medium, its dynamic state is characterized by the time-dependent population inversion, $N(\rho_{22} - \rho_{11}) \equiv [N_2(t) - N_1(t)]$, and the complex amplitude of the atomic coherence, $\tilde{\rho}_{21}(t) = \tilde{\rho}_{21}^*(t)$, defined in (11.34a) and (11.34b). The corresponding rate equations are:

$$\frac{\mathrm{d}}{\mathrm{d}t}(N_2 - N_1) + \frac{1}{T_1}\left[(N_2 - N_1) - (N_2^{(0)} - N_1^{(0)})\right]$$
$$\cong \ \mathrm{i}\frac{2e\mathrm{N}}{\hbar}\left[\tilde{E}_z^*(t)z_{12}\tilde{\rho}_{21} - \tilde{E}_z(t)z_{21}\tilde{\rho}_{12}\right], \tag{11.56a}$$

$$\left[\frac{\mathrm{d}}{\mathrm{d}t} + \mathrm{i}(\omega_{21} - \omega_0) + \frac{1}{T_2}\right]N\tilde{\rho}_{21} \cong \frac{\mathrm{i}}{\hbar}(N_2 - N_1)\,ez_{21}\tilde{E}_z(t). \tag{11.56b}$$

They are similar to (11.34a) and (11.34b), except here the population inversion and the complex amplitude of the atomic coherence are now time-dependent. The E-fields in these equations are treated as classical variables. For laser applications, because the light intensity is generally high and the photon number is always very large, as shown in Section 5.4, the field can be treated classically, unless noise characteristics of the laser or the statistical properties of the laser output beam are being considered.

To describe the dynamics of the laser, the complex amplitude of the E-field in these equations is the intra-cavity time-dependent field of the laser, which must satisfy the corresponding Maxwell's equation and the suitable laser cavity boundary conditions on the field. For a qualitative understanding of some of the basic features of the dynamic properties of lasers in general, one can simplify the problem by characterizing the time-variation of the electromagnetic field inside the cavity in terms of the intensity of the field only, rather than the intensity and the phase of the E-field. Thus, the time rate of change of the intracavity intensity of the light consists of a cavity-loss term and a stimulated emission term as given by (11.48), and, using the notation of (8.32), is:

$$\frac{d}{dt}|\tilde{E}_z(t)|^2 = -\frac{1}{T_{ph}}|\tilde{E}_z(t)|^2 + [N_2(t) - N_1(t)]\,\sigma_{st}c|\tilde{E}_z(t)|^2 + 0(|\tilde{E}_z{}^{\text{spont}}(t)|^2).$$

(11.57)

The last term represents a small amount of noise to account for the spontaneous emission by the laser medium into the lasing mode of the cavity. This small term is always implicitly present at the start in the equation in order to initiate the lasing action when there is positive gain in the laser medium, but is considered negligible once the intracavity laser intensity is above the noise level.

If T_2 is much shorter than the population relaxation time and T_1, there is a second powerful approximation that can be made. It is known as the "adiabatic approximation," which assumes that the transient part of the solution of Eq. (11.56b) can be neglected, and that the time-dependent complex amplitude of the atomic coherence, $\hat{\rho}_{21}(t)$, follows adiabatically the population difference:

$$\tilde{\rho}_{21}(t) \cong \frac{[\rho_{22}(t) - \rho_{11}(t)]}{\hbar}\left[\frac{e\,z_{21}\tilde{E}_z(t)}{(\omega_{21} - \omega_0) - iT_{21}^{-1}}\right].$$

(11.58)

This approximation eliminates Eq. (11.56b) as an independent equation by keeping only (11.56a) for the medium and (11.57) for the field. Substituting (11.58) into (11.56a) gives the rate equation for the population difference:

$$\frac{d}{dt}(N_2 - N_1) \cong -\frac{N_2 - N_1}{T_1} + \frac{N_2^{(0)} - N_1^{(0)}}{T_1} - |\tilde{E}_z|^2(N_2 - N_1)\frac{\sigma_{st}c}{\pi\hbar\omega_0}.$$

(11.59)

The three terms in (11.59) show that the rate of increase of the population inversion consists of: first, a negative term corresponding to the population relaxation process; second, a positive term corresponding to a steady-state population inversion in the absence of any radiation; and third, a negative term corresponding to the stimulated emission process. To maintain a positive population inversion as is required for lasing action, the second term must be positive and maintained by an external "pumping mechanism." It is common practice to represent this contribution to the increase in the population inversion by a phenomenological "pumping rate" R_{pump} explicitly instead of the term $(N_2^{(0)} - N_1^{(0)})/T_1$. Thus, the final set of "laser rate equations" for the population inversion $(N_2 - N_1)$ and photon density $N_{ph}(t) \equiv \frac{|\bar{E}_z(t)|^2}{2\pi h \nu_0}$ is of the form:

$$\begin{cases} \frac{d}{dt}(N_2 - N_1) + \dfrac{N_2 - N_1}{T_1} = -2Bg_f(\nu_0)h\nu_0(N_2 - N_1)N_{ph} + R_{pump}, \\ \frac{d}{dt}N_{ph} = -\dfrac{1}{T_{ph}}N_{ph} + Bg_f(\nu_0)h\nu_0\left[N_2(t) - N_1(t)\right]N_{ph} + o(N_{ph}^{(spont)}), \end{cases} \quad (11.60)$$

where $B \equiv \sigma_{st}c/g_f(\nu_0)h\nu_0 = 2\pi e^2|x_{12}|^2/\hbar^2$ is the well-known "Einstein B-coefficient" for stimulated emission, as defined in (8.32), and $g_f(\nu_0)$ is the line shape function defined in (11.45).

Note that, because of the coupling term $(N_2 - N_1)N_{ph}$ between the two equations, this set of equations, (11.60), is nonlinear. In general, it can only be solved numerically. Let us now consider two numerical examples using numbers more-or-less typical of (a) a gas laser: $T_1 \sim 10^{-8}$, $T_2 \sim 10^{-9}$, $T_{ph} \sim 10^{-7}$, $Bg_f(\nu_0)h\nu_0 = \sigma_{st}c \sim 10^{-5}\mathrm{cm}^3\mathrm{sec}^{-1}$, $R_{pump}(t>0) \sim 10^{21}\,\mathrm{cm}^{-3}\mathrm{s}^{-1}$, $[N_2(t=0) - N_1(t=0)] = 0$, and $N_{ph}(t=0) = 0^+$ which is an arbitrarily chosen very small number; and (b) a solid-state laser: $T_1 \sim 3 \times 10^{-4}$, $T_2 \sim 10^{-12}$, $T_{ph} \sim 10^{-8}$, $Bg_f h\nu_0 = \sigma_{st}c \sim 10^{-8}\mathrm{cm}^3\mathrm{sec}^{-1}$, $R_{pump}(t>0) \sim 5 \times 10^{22}$, $[N_2(t=0) - N_1(t=0)] = 0$, and $N_{ph}(t=0) = 0^+$. In the laser literature, the term "solid state lasers" generally refers to lasers with active media that are insulating crystals doped with impurity ions, such as rare earth or transition-metal ions (e.g. Nd^{3+} doped yttrium aluminum garnet crystal, or ruby crystal which is Cr^{3+} doped Al_2O_3). Semiconductors are, of course, solids also, but such lasers are usually identified explicitly as "semiconductor lasers." Because of the short atomic coherence times T_2 in almost all lasers, the rate equations (11.60) are applicable in the vast majority of practical cases. The calculated transients of the two types of lasers after the pump is abruptly turned on from $R_{pump}(t \le 0) = 0$ to $R_{pump}(t \ge 0) > 0$ at $t = 0$ are shown in Figure 11.2(a) and (b).

It is interesting to note the significant qualitative difference between the two. In the case of the gas lasers, the T_1 time is generally much shorter than the photon life time T_{ph}. As a result, once the field in the cavity builds up due to stimulated emission, the population inversion can quickly follow the time-varying electromagnetic energy in the cavity adiabatically. There is no transient oscillation of the population inversion or the intracavity electromagnetic energy around their respective steady-state values. The population inversion reduces adiabatically as the laser emission builds up smoothly to the final steady-state value, as shown in Figure 11.2(a).

In the case of the solid-state lasers, both the population inversion and the laser radiation intensity show strong transient oscillations. They are known as the laser "relaxation oscillations." In this case, the population relaxation time T_1 is in general much longer than the photon life time T_{ph}. This allows a substantial fraction of the energy pumped into the laser medium to be stored in the population inversion initially, even as the electromagnetic field in the cavity builds up due to stimulated emission. As a result, the population inversion can far exceed the threshold population inversion

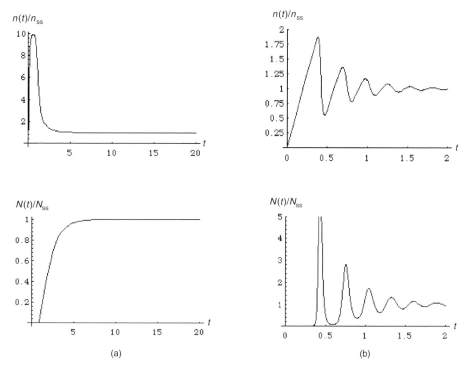

Figure 11.2. Numerical examples of the turn-on transient dynamics of two types of lasers based on the solutions of the density-matrix equations. (a) is typical of lasers where the cavity-photon lifetime T_{ph} is much longer than the population relaxation time T_1 in the laser medium, such as in a gas laser. (b) is typical of lasers where the population relaxation time T_1 is much longer than the cavity-photon lifetime T_{ph}, such as in an insulating solid-state laser. (t in 10^{-7} s, $n \equiv N_2 - N_1$, $N_{ss} \equiv$ steady-state, $N \equiv N_{ph}$.)

needed to initiate lasing action in the cavity. Once the stimulated emission starts, the intracavity photons can build up quickly and exceed the final steady-state value also, but the population inversion cannot adjust and reduce fast enough due to the long relaxation time. This rapid increase in the field can lead to a stimulated emission rate far exceeding the pump rate and eventually reduce the population inversion to below the final steady-state value. As this happens, the field intensity and, consequently, the rate of stimulated emission will start to decrease to the point where it becomes less than the pumping rate. It will then in turn lead to a build-up of the population inversion again. Such cycles of rapid build-up followed by rapid decrease of the electromagnetic fields in the cavity and the corresponding delayed response of the population inversion will repeat until the transients slowly die off, as illustrated by the numerical results in Figure 11.2(b). This behavior is typical of solid-state lasers, where the population relaxation time T_1 is generally much longer than the photon life time T_{ph}. The relaxation oscillation period is typically approximately equal to the geometric mean of the population relaxation time and the photon lifetime, and damps out in a time somewhat faster than the population relaxation time. Lasers that are characterized by strong turn-on transients in the form of sharp relaxation oscillations also tend to be potentially unstable and can exhibit a rich variety of chaotic and spiky types of dynamic behavior.

This is just one simple example of the dynamic characteristics of nonlinear-coupled quantum systems basic to laser physics that can be understood on the basis of the density-matrix equations.

11.5   Problems

11.1   Consider a medium consisting of a statistical ensemble of N spin-1/2 particles per volume. The matrices representing the Cartesian components of the spin angular momentum of such particles in the representation in which $\hat{S}_z$ and $\hat{S}^2$ are diagonal are given in (6.50). Give the averaged expectation values per volume of the three components of the spin angular momentum in terms of the appropriate density matrix elements for the statistical ensemble of particles.

11.2   An electrically charged spinning particle with a spin angular momentum will have a magnetization proportional to the spin angular momentum. Suppose the averaged expectation value of the magnetization of the medium considered in Problem 11.1 is $\vec{M} = N \, \text{trace} \, [\hat{\rho}(\gamma \vec{S})]$.

(a)   Express the three Cartesian components of the magnetization in terms of the appropriate density-matrix elements as in Problem 11.1.

(b)   Write the Hamiltonian of the spin-1/2 particles in the presence of a static magnetic field $\vec{H} = H_x \vec{x} + H_y \vec{y} + H_z \vec{z}$, but in the absence of any relaxation processes. Making use of the results of part (a), show on the basis of the density-matrix equation (11.16) that the dynamic equation describing the precession of the magnetization $\vec{M}$ around such a magnetic field is:

$$\frac{d\vec{M}}{dt} = \gamma \vec{M} \times \vec{H},$$

just like in classical mechanics.

(c)   Suppose a magnetic field consisting of a static component in the $\vec{z}$ direction and a weak oscillating component in the plane perpendicular to the $\vec{z}$ axis is applied to the medium: $\vec{H} = H_0 \vec{z} + \vec{H}_x \, \vec{x} \equiv H_0 \vec{z} + H_1 \cos \omega_0 t \, \vec{x}$. Show on the basis of the corresponding density-matrix equations (11.27) that the equations of motion for the three components of the magnetization are of the form:

$$\frac{dM_z}{dt} = -\frac{M_z - M_z^{(\text{th})}}{T_1} + i\frac{\gamma}{2} H_1 (\cos \omega_0 t)(M_- - M_+),$$

$$\frac{dM_\pm}{dt} = -\frac{M_\pm}{T_2} \pm i\gamma H_0 M_\mp \mp i\gamma H_1 (\cos \omega_0 t) M_z,$$

where $M_\pm = M_x \pm i M_y$. These are the well-known Bloch equations in the literature on magnetic resonance phenomena.

11.3 (a) Show that the intensity attenuation coefficient for a monochromatic elec-
        tromagnetic wave propagating in an optical medium of two-level atoms
        given in (11.44) can also be put in the form:

$$\alpha_z = \frac{\varepsilon_{zz}'' \omega_0}{\sqrt{\varepsilon_0} c} = \frac{\omega_p^2 f}{4\sqrt{\varepsilon_0} c} \, g(\nu_0),$$

        where $\omega_p \equiv \sqrt{\frac{4\pi(N_1 - N_2)e^2}{m}} \approx \sqrt{\frac{4\pi Ne^2}{m}}$, if most of atoms are in the ground state,
        is known as the "plasma frequency" and $f_z \equiv \frac{2m\omega_{21}}{\hbar} |z_{21}|^2$ is known as the
        "oscillator strength."
     (b) Compare the result obtained in part (a) with the classical result based on a
        damped harmonic oscillator model instead of the two-level atom model
        Suppose the equation of motion of the harmonic oscillator is of the form:

$$\frac{d^2}{dt^2} z(t) + \Gamma \frac{d}{dt} z(t) + \omega_{21}^2 z(t) = -\frac{f^{1/2}e}{m} \left(\tilde{E}_z e^{-i\omega_0 t} + \tilde{E}_z^* e^{i\omega_0 t} \right),$$

        which describes the oscillating motion of a particle of mass m and negative
        charge of magnitude $f^{1/2}e$ bound to a fixed point in space, similar to the
        oscillator shown in Figure 5.1. The spring constant of the harmonic oscilla-
        tor is equal to $m\omega_{21}^2$; the damping constant is Γ; and the deviation of the
        particle from its equilibrium position in the absence of any electric field
        E_z is $z(t)$. For the classical result, assume $\omega_0 \approx \omega_{21} \gg \Gamma^{-1}$ so that
        $\omega_0^2 - \omega_{21}^2 \cong 2\omega_0(\omega_0 - \omega_{21})$.
     (c) Make a similar comparison as in part (b) of the real parts of the complex
        dielectric constants. Based on these results and those of part (b), discuss
        the physical interpretation of the concept of "oscillator strength f" defined
        above.

11.4 Substitute (11.51) into (11.27) and show that it is indeed a solution of the
     density-matrix equation (11.27).

11.5 Show that the second-order nonlinear optical susceptibility $\tilde{\tilde{\chi}}^{(2)}(\omega_1 + \omega_2 = \omega_3)$,
     which is a third-rank tensor, must vanish for any optical medium with inversion
     symmetry.

11.6 Consider a semiconductor laser with the following parameters: $T_1 \sim 10^{-9}$, $T_2 \sim$
     10^{-12}, $T_{ph} \sim 5 \times 10^{-12}$, $R_{pump} \sim 10^{27} \text{cm}^{-3}\text{s}^{-1}$, $Bh\nu_0 g_f$ $(\nu_0) \sim 6 \times 10^{-7} \text{cm}^{-3}\text{s}^{-1}$.
     Calculate numerically (using, for example, 'Mathematica') the turn-on
     dynamics of such a laser using the laser-rate equations as in the numerical
     examples shown in Figure 11.2.

References

Ballhausen, C. J. and Gray, H. B. *Molecular Orbital Theory*. New York: W. A. Benjamin, Inc., 1964.

Bethe, H. A. and Jackiw, R. *Intermediate Quantum Mechanics*. 3rd edn. Menlo Park, CA: The Benjamin Publishing Company, 1986.

Bohm, D. *Quantum Theory*. New York: Prentice Hall, 1951.

Cohen-Tannoudji, C. Diu, B. and Laloë, F. *Quantum Mechanics*, Vols. I and II. New York: John Wiley & Sons, 1977.

Condon, E. U. and Shortley, G. H. *The Theory of Atomic Spectra*. Cambridge: Cambridge University Press, 1963.

Coulson, C. A. *Valence*. 2nd edn. London: Oxford University Press, 1961.

Dirac, P. A. M. *Quantum Mechanics*. 3rd edn. London: Oxford University Press, 1947.

Edmonds, A. R. *Angular Momentum in Quantum Mechanics*. Princeton, NJ: Princeton University Press, 1957.

Glauber, R. J. Coherent and incoherent states of the radiation field. *Phys. Rev.* **131** (1963), 2766–2788; *Phys. Rev. Letters* **10**, 84–86.

Gray, H. B. *Chemical Bonds: an Introduction to Atomic and Molecular Structures*. Menlo Park, CA: Benjamin/Cummings Publishing Company, 1973.

Heitler, W. *The Quantum Theory of Radiation*. 3rd edn, London: Oxford University Press, 1954.

Herzberg, G. *Atomic Spectra and Atomic Structure*. New York: Dover Publications, 1944.

Kittel, C. *Introduction to Solid State Physics*, 7th edn. New York: John Wiley & Sons, 1996.

Pauling, L. *The Chemical Bond*. Ithaca, NY: Cornell University Press, 1967.

Rose, M. E. *Elementary Theory of Angular Momentum*. New York: John Wiley & sons, 1956.

Shen, Y. R. *Principles of Nonlinear Optics*. New York: John Wiley & Sons, 1984.

Siegman, A. E. *Lasers*. Mill Valley, CA: University Science Books, 1986.

Smith, R. A. *Semiconductors*. Cambridge: Cambridge University Press, 1964.

Streetman, B. G. *Solid State Electronics*, 4th edn. Englewood, NJ: Prentice Hall, 1995.

Yariv, A. *Quantum Electronics*, 3rd edn. New York: John Wiley & Sons, 1989.

Index